Japanese Commons in the Coastal Seas

Tetsuo Yanagi

Japanese Commons in the Coastal Seas

How the Satoumi Concept Harmonizes Human Activity in Coastal Seas with High Productivity and Diversity

Springer

Tetsuo Yanagi
Professor, Doctor of Science
Research Institute for Applied Mechanics
Kyushu University
Fukuoka, Japan

Original Japanese edition published by KOUSEISHA KOUSEIKAKU Co., Ltd.
Satoumi Seouseiron

ISBN 978-4-431-54099-1 ISBN 978-4-431-54100-4 (eBook)
DOI 10.1007/978-4-431-54100-4
Springer Tokyo Heidelberg New York Dordrecht London

Library of Congress Control Number: 2012941502

Printed on acid-free paper

Springer is part of Springer Science+Business Media (www.springer.com)

Preface

I first proposed the term “satoumi” in 1998 to refer to the concept of a coastal sea with high biological productivity and high biodiversity due to harmonized human activities. Since then I have received many questions, comments, objections, and other reactions to this new concept for coastal sea management. I have proceeded with my study of the satoumi concept to answer such objections and comments. At the same time, new trials for creating satoumi have begun in other locations—not only in Japan but also in Indonesia.

I published a book in Japanese about “Satoumi Creation” in 2010 in order to introduce the development of the satoumi concept and to provide examples of satoumi creation to the Japanese people. I now have translated that book into English because I believe the satoumi concept will be useful not only in Japan but also in other countries, where many people suffer from degradation of the coastal environment.

I welcome comments from overseas about this book, because they will be useful for further development of the satoumi concept and successful coastal sea management around the world.

Finally, I express my appreciation to the editorial staff members of Springer Japan for their kind assistance in the publication of this book.

Fukuoka, Japan
January 2012

T. Yanagi

Contents

Chapter 1
Introduction

I proposed the Satoumi concept, as an analogy of the Satoyama concept, as "Coastal sea with high biodiversity and productivity under the human interaction" in 1998 (Yanagi 1998a, b), and published a book in English on Satoumi (Yanagi 2007a). Since that time, this new concept for environmental conservation in the coastal sea has been widely accepted and was included in the Japanese national policy of "Strategy for Establishment of Environmental Nation" in 2007.

At the same time, many people asked me many questions on the new and not-yet-mature concept of Satoumi. For example: "Does biodiversity increase by human interaction in the coastal sea?" "You have to consider the economic aspect of fishing villages in detail." "Some legal support is necessary for the creation of Satoumi." "How about the relationship between God and Nature in Satoumi?" "How about the relationship between fishermen and city people?" and so on.

This book is written to reply to some of these questions. In Chap. 2, the relationship between biodiversity and human interaction, economic problems related to Satoumi, legal support for the creation of Satoumi, the landscape ecology point of view of Satoumi, and the relationship between society and science related to the Satoumi movement are discussed. In Chap. 3, some examples of the creation of Satoumi in Japan are introduced. In Chap. 4, the overseas appeal of the Satoumi concept are introduced, and the conclusion is presented in Chap. 5.

T. Yanagi, *Japanese Commons in the Coastal Seas: How the Satoumi Concept Harmonizes Human Activity in Coastal Seas with High Productivity and Diversity*, DOI 10.1007/978-4-431-54100-4_1, © Springer Japan 2013

Chapter 2
Development of the Satoumi Concept

2.1 Biodiversity and Human Interaction

The new concept of Satoumi, which means "Coastal sea with high biodiversity and productivity under human interaction," has been widely accepted and is included in the Japanese national policy such as "Strategy for Establishment of Environmental Nation in 21st Century" (2007.6), "3rd National Strategy for Bio-diversity" (2007.11), "Basic Marine Plan" (2008.3), and "National Economic Growth Strategy" (2008.6). The Environmental Agency, Japan began to support the activities of the creation of Satoumi in 2008.

It is easy for people to understand that human interaction increases biodiversity at Satoyama (e.g., Morimoto 2008). On the other hand, there is no scientific evidence on the relationship between human interaction and biodiversity in the coastal sea, and some people say that human interaction decreases biodiversity in the coastal sea. This concern may arise because we have had many examples in the past, such as land reclamation and seawall construction, that have destroyed the marine ecosystem in the coastal sea. I introduce some examples that show the increase of biodiversity and productivity under human interactions in the coastal sea in this section.

The terminology of biodiversity has three meanings: (1) ecosystem diversity, (2) species diversity, and (3) genetic diversity. I use the terminology of biodiversity to mean species diversity in this book.

T. Yanagi, *Japanese Commons in the Coastal Seas: How the Satoumi Concept Harmonizes Human Activity in Coastal Seas with High Productivity and Diversity*, DOI 10.1007/978-4-431-54100-4_2, © Springer Japan 2013

2.1.1 Human Interaction to Not Decrease Biodiversity: Extermination of Starfish in the Coral Reef and Herbivorous Fish in the Marine Forest

Some divers exterminate starfish in the coral reefs at Okinawa or off Kushimoto to preserve the coral reefs (Kakuma 2009b). Biodiversity in the coral reefs is maintained by such human actions of extermination.

The fishermen in Shonan area, Shizuoka Prefecture, succeeded in the rehabilitation of the marine forest within a 1,500 m^2 area by setting blocks with sea algae seeds and the extermination of herbivorous fish using a standing net in 2000 (Hagiwara 2008; details are shown in Sect. 3.3). Biodiversity with abalone, turbo, sea urchins, shrimps, crab, sea anemones, and other organisms that live in the marine forest can be sustained by such human interactions.

These examples show that human interaction can maintain high biodiversity in the coastal sea.

2.1.2 Human Interaction to Increase Biodiversity and Productivity: Ishihimi (Rock Tidal Weir)

There is a kind of set net called Ishihimi that is made of rocks in Japan (Ariake Bay, Okinawa), Korea, and Taiwan (Tawa 2006). Ishihimi is a structure with a half-circle shape resembling a horse's hoof on tidal flats or coral reefs that have large tidal amplitude which can trap fish at ebb tide, because fish that enter at the flood tide cannot escape from Ishihimi at ebb tide (Fig. 2.1a–c). Such a structure is called Nagaki or Kachi in Okinawa, Japan (Kamimura 2007).

Biodiversity increases around Ishihimi because some attached biota can survive on the surface of the accumulated rocks; some grazers of these attached biota will then come to feed, and other sea animals come to use the rocks for a hiding place. Therefore, biodiversity at an area with Ishihimi is different from that in areas without Ishihimi. Ishihimi is a kind of habitat or biotope made by mankind.

High biodiversity is possible under the conditions of many kinds of environmental habitats for biota. Ishihimi, caused by human interaction, created a new environment for biota that resulted in high biodiversity.

Quantitative data on the difference of biodiversity in the area with Ishihimi and without Ishihimi have been obtained, and biodiversity has increased after the construction of Ishihimi (Fig. 2.2; Kamimura 2011). Comparison of biodiversity of benthos in areas with and without an artificial reef is introduced in Sect. 2.8.1.

Ishihimi has disappeared now because of the development of the modern net trap (Tawa 2006), but it is a good example that suitable human interaction can increase biodiversity and productivity in the coastal sea at the same time. Tawa (2006) wrote that there is no Ishihimi in the Seto Inland Sea, Japan, but Ishihimi

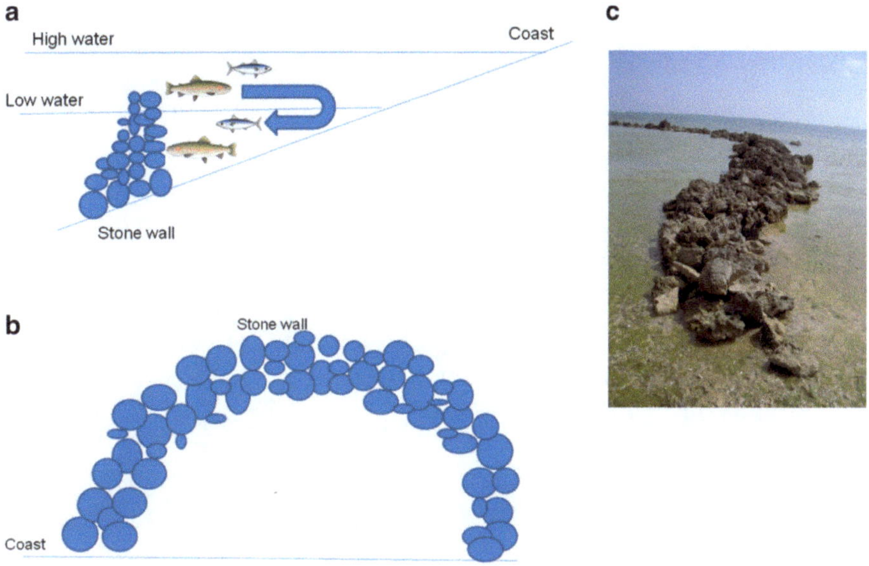

Fig. 2.1 (**a**) Vertical view of Ishihimi. (**b**) Plan view of Ishihimi. (**c**) Ishihimi at Shiraho in Ishigaki Island, Okinawa

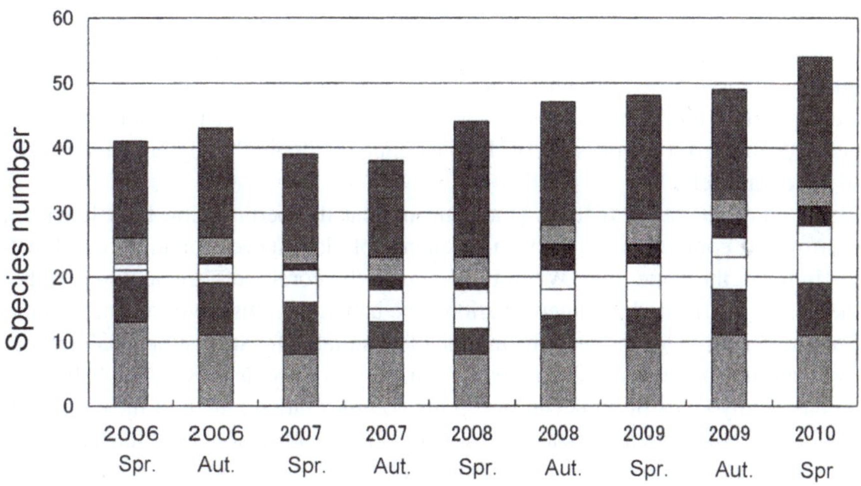

Fig. 2.2 Change of biodiversity before and after the construction of Ishihimi at Shiraho in Autumn, 2006. *Spr* spring, *Aut* autumn Colors denote different species (from Kamimura 2011)

existed until 40 years ago in the southern part of Suo-Nada in the Seto Inland Sea (Fig. 2.3). An Ishihimi for sightseeing and environmental education is being rebuilt now there.

Fig. 2.3 Ishihimi at Suo-Nada in the Seto Inland Sea, Japan

2.1.3 Biodiversity and Hurricanes in the Coral Reef

Connell (1978) investigated the relationship among hurricane intensity, its frequency, elapsed time after hurricane passing, and biodiversity in coral reefs, and clarified that moderate intensity and frequency of hurricanes and moderate elapsed time from the passage of the hurricane results in the highest biodiversity in the coral reefs.

We can understand that his study means that the perturbation by a hurricane destroys the coral reefs and stops the sequence of plant succession in the coral reefs to climax at the same time. When the plants in the coral reefs are at climax (when there is no hurricane-based perturbation), the habitats in this environment for biota become very simple, and only the marine biota suitable for such simple habitats can live, and thus biodiversity decreases. When the hurricane destroys some habitats at climax, many kinds of biota can live in the complex environment of the coral reefs and biodiversity increases. The effects of a moderate hurricane cause the habitats of the coral reef complex to be at subclimax and thus of high biodiversity. A hurricane is a natural perturbation, but human perturbations such as cleaning weeds or planting woods in Satoyama play a role similar to that of a hurricane.

We can see a similar situation in seagrass beds in the coastal sea: that is, many kinds of small fish gather at the rim or in the gaps of the seagrass bed, as observed by scuba diving. We carried out field observations on the difference of biodiversity between the central part of the seagrass bed and its open spaces in Mitsukuchi Bay, the Seto Inland Sea, Japan (Fig. 2.4). When we make suitable gaps in the seagrass beds by human interaction, biodiversity in the seagrass beds will increase.

Fig. 2.4 Fish and eelgrass beds (modified from Tanimoto 2010)

2.1.4 Biodiversity and Eutrophication in Tidal Flats

Kokubu and Takayama (2008) investigated the relationship between chemical oxygen demand (COD) value and individual number of benthos at the tidal flats in Ago Bay, Mie Prefecture, and clarified that the number of individual benthos becomes maximum at a medium COD value. They say that it is possible to artificially increase biodiversity at the tidal flats by controlling the COD value there. That is, increasing the COD value by adding dredged bottom mud to the low-COD tidal flats and decreasing COD by adding clean sand to the high-COD tidal flats are good practices for increasing biodiversity there.

They observed the individual number and species number of benthos at tidal flats with different COD values in Ago Bay, Japan, and clarified that biodiversity is increased by controlling the COD value at tidal flats.

The foregoing discussions are summarized in Fig. 2.5; that is, a large or small perturbation results in low biodiversity, but moderate perturbations result in the highest biodiversity.

2.1.5 Relationship of Human Interaction and Biodiversity in Satoumi

The relationship between human interaction and biodiversity in Satoumi is summarized in Fig. 2.6. One axis shows the human creation of habitats such as Ishihimi, and another axis shows the human interaction of stopping the transition

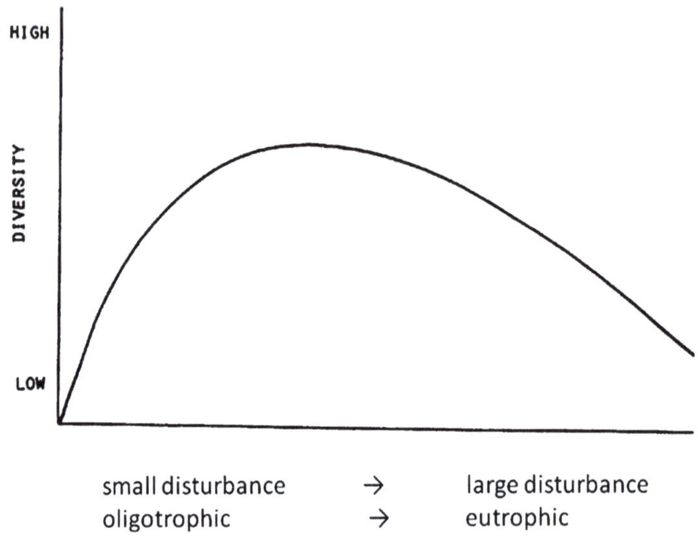

Fig. 2.5 Relationship between biodiversity and disturbance and eutrophication

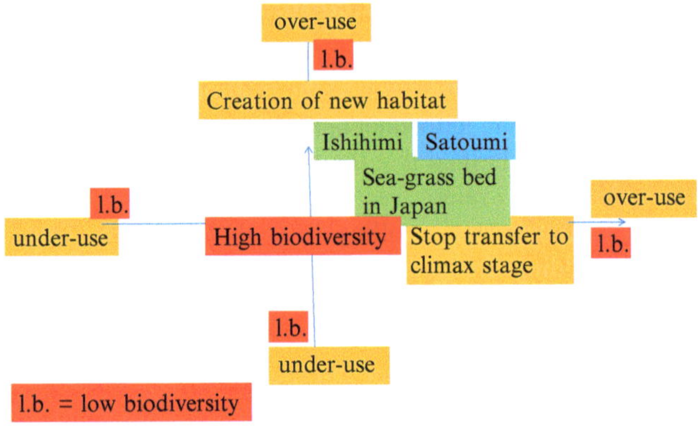

Fig. 2.6 Relationship between human interaction and biodiversity

to climax of flora such as eelgrass beds. Along both axes, underuse and overuse result in low biodiversity, and only moderate interactions (wise use) can result in high biodiversity.

In Satoyama, human interaction mainly aims to stop plant succession from reaching its climax stage; that is, the deciduous broadleaf forest at Satoyma in the western part of Japan will evolve into evergreen forest, which are the primary trees there, in the long term without human interaction. There is little sunlight at the bottom of an evergreen leaf forest, and biodiversity there is much lower compared to that in the deciduous broadleaf forest.

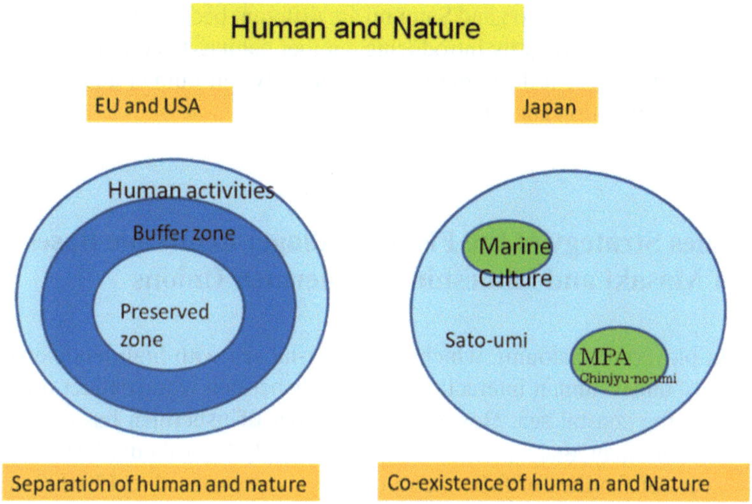

Fig. 2.7 Relationship between nature and human in Western countries (*left*) and Japan (*right*)

On the other hand, human interactions at Satoumi mainly aim to conserve many kinds of environmental habitats or to create new habitats or a new environment for marine biota; also, sometimes people cut eelgrass from the beds to obtain fertilizer.

2.1.6 Different Relationships Between Humans and Nature in Europe and Japan

There is a big difference in the thinking about the relationship between humans and nature in Western countries and in Japan. In Western countries, human activity and nature are completely separated, and a buffer zone is formed between them to preserve the natural world. On the other hand, in Japan human activity and nature always coexist, and human activities to conserve nature are a principal focus of human wisdom (Fig. 2.7) because the high population density in Japan means that the Japanese cannot preserve separate zones for nature.

We cannot say easily which attitude is better. However, some people in Western countries are beginning to say that we should consider the beauty of the style of thought of Satoyama and Satoumi (International EMECS Center 2009).

Increasing biodiversity in Satoumi does not mean increasing local biodiversity: for example, building a wave-breaking wall at a sand beach with its resulting attached biomass is not recommended because building such a wall destroys the whole sand beach ecosystem. We have to consider the conservation of the ecosystem in the entire area and suitable human interactions to increase the biodiversity of the whole ecosystem.

The concepts of mitigation and Satoumi are different because mitigation aims for zero net loss of biodiversity by introducing another habitat. In contrast, the Satoumi concept aims to create a desirable relationship between human and nature, even including the mitigation concept.

2.2 Sales Strategy of the Fishing Village: Economic Bases of Masaki and Himeshima Fishermen Unions

The main players in Satoumi, which is the coastal sea with high biodiversity and productivity under human interactions, are the fishermen who conduct production activities in the coastal sea. However, the number of fishermen has decreased in recent years in Japan, that is, from about 860,000 in 1946 to about 260,000 in 2004. The numbers of families of fishermen in Japan also decreased to 130,000 in 2004, that is, minus 2% from 2003 (Fisheries Agency 2006). Such decrease of fishermen affects not only Satoumi creation activities but also the maintenance of the multiple functions of fisheries, that is, the material cycling function, environmental conservation function, and ecosystem conservation function (Matsuda 2007).

It is needless to say that the economic aspect is the basis for the maintenance of fishing villages and the life of fishermen.

The economic bases and activities of two active fishermen unions, Misaki Fishermen Union in Ehime Prefecture (Yanagi 2005b) and Himeshima Fishermen Union in Ooita Prefecture (Yanagi 2004), are introduced in this chapter to point out the important aspects for maintenance of fishing villages and the creation of Satoumi.

2.2.1 Misaki Fishermen Union

The full membership of Misaki Fishermen Union, which is located at the tip of Misaki Peninsula, Ehime Prefecture, is 254, and all harvests are on sale through the Union (Yanagi 2005b). The yearly total sale of Misaki Fishermen Union was about 10 billion Japanese yen in the fiscal year of 2005; sales through the market were only 54%, and sales by direct contact and direct shops were 46% (Table 2.1).

The reason why there are no sales through the local market is that the local economy has not sufficient power to buy the harvest of Misaki Fishermen Union and the Union finds it is better to sell their harvest directly at large markets such as Tokyo or Osaka to avoid the intermediate margin. Therefore, there has been no sale through the local market since 1985, and the harvests are sold directly to each district in Japan depending on the favorites of the people in that district: for example, horse mackerel to Kanto district, abalone to Kansai district, and swordfish to Kyushu district.

Table 2.1 Sales of Misaki Fishermen Union in 2005

	Hokkaido	%	Tohoku	%	Kanto, Shin-etsu	%	Chubu, Hokuriku	%	Kansai	%	Chugoku	%	Shikoku	%	Kyushu	%	Local	%	Sum	%
Market	0	0	12,244,404	2	13,334,151	2	9,123,621	1	118,553,096	21	21,726,757	4	316,810,912	57	72,285,685	13	0	0	564,078,626	54
Direct send	10,710	1	0	0	91,892,922	62	4,741,643	3	11,895,731	8	3,015,916	2	34,478,263	23	986,685	1	0	0	147,021,870	14
Direct shop	376,142	1	229,527	1	19,082,911	27	1,970,772	2	5,898,359	8	13,710,421	18	21,598,013	30	1,328,206	1	8,987,648	12	73,181,999	7
Direct sell	0	0	0	0	3,784,747	1	111,295	1	1,347,863	1	0	0	2,525,766	30	317,715	1	260,505,750	95	268,593,136	25
Sum	386,852	1	12,473,931	1	128,094,731	12	15,947,331	1	137,695,049	13	38,453,094	3	375,412,954	36	74,918,291	7	269,493,398	26	1,052,875,631	

Table 2.2 Sales of Himeshima Fishermen Union in 2006

	Yen	%
Line fishing	339,004,812	40.0
Drift net	202,168,178	23.7
Set net	105,417,638	12.4
Hole fishing	49,993,347	5.9
Longline fishing	45,806,075	5.4
Dive fishing	33,531,912	3.9
Collecting	25,190,042	3.0
Winding net	22,082,841	2.6
Others	28,649,095	3.4
Sum	851,843,940	100

The targets of direct sale are hotels and departments in big cities such as Tokyo and Osaka, which occupies 14% of total sales. The Union has two direct shops at Misaki and Matsuyama, which provides 7% of total sales. Direct sales through the Internet account for 25% of total sales.

The necessary cost for the sales generating 1.7 billion Japanese yen is supported by the Union from the total sale income of 10.5 billion Japanese yen, and the remaining income of 9.8 billion Japanese yen is distributed to the members of the Union. The income of 100 fishermen among the full membership, who work as divers for half the year and as orange farmers in the other half year, is 2.3 billion Japanese yen, which is a quarter of the total income of the Union and 2,300,000 yen for each member. The necessary cost of diving is only 1%, and their real income is 2,280,000 yen in a year. The remaining 150 fishermen, who work at line fishing, receive 4,700,000 yen per year, but the necessary cost for line fishing is 14% and thus their real income is 4,000,000 yen per year (330,000 yen per month).

Table 2.1 shows that the sale at Shikoku of 36% is larger than the local market of 26% and that at Kansai (mainly Osaka) is 13%, and the sale at Kanto (mainly Tokyo) of 12% is next. Such facts suggest that the big cities such as Osaka and Tokyo support the economy of the local fishing village of Misaki.

The sales leader, Mr. Sonoda, of the Union says that "We want to increase the ratio of direct selling but we have to sustain 40% sale through the market because we must be sensitive to the favorites of the usual costumers through the market." He stresses that too much dependence on direct sales decreases the sensitivity of fishermen to provide good fishing products for the favorites of the usual customers.

2.2.2 *Himeshima Fishermen Union*

The number of full members of Himeshima Fishermen Union, which is located off Kunisaki Peninsula in the northern part of Ooita Prefecture, is 190, and the union's sales in fiscal year of 2006 was 8.5 billion Japanese yen. Its content is shown in Table 2.2.

Line fishing for swordfish occupies 40%, drift nets for shrimp 24%, set nets for flatfish 12%, and hole fishing for octopus, longline fishing for balloon fish, and diving for abalone follow in importance.

All the harvest of Himeshima Fishermen Union is sold through the local market. Brokers sell 100% of red sea bream and 30% of abalone to Osaka and 100% of shrimp to Tokyo.

The average income of a full member of the Union was about 4,500,000 yen per year, but as the necessary cost for line fishing and net fishing was about 30%, the real income was 3,100,000 Japanese yen per year (260,000 Japanese yen per month). The full member must pay 4% of his income to the Union.

The Himeshima Fishermen Union wants to carry out direct selling but it is not easy to develop the sales route. The leader of the Union, Mr. Kitamura, says that "When the harvest is too much we cannot sell all the harvest by direct selling. Therefore the local broker is necessary."

2.2.3 Comparison of Misaki and Himeshima Fishermen Unions

The yearly income of the full member of Misaki Fishermen Union of 4,000,000 Japanese yen (330,000 yen per month) and that of a Himeshima member of 3,100,000 Japanese yen (260,000 yen per month) are higher than the average income of Japanese fisherman of 2,150,000 Japanese yen (Fisheries Agency 2006) in 2004. We show why the income of Misaki Fishermen Union is 1.3 times higher than that of Himeshima Fishermen Union in this section.

The catch and sale of horse mackerel by line fishing of Misaki Fishermen Union in 2005 were 58,000 kg and 96,716,000 Japanese yen, respectively, and the unit price of horse mackerel was 1,668 Japanese yen/kg. On the other hand, catch and sales by set net of Himeshima Fishermen Union in 2006 were 15,409 kg and 96,716,000 yen, respectively, and the unit price was 410 Japanese yen/kg. The unit price of horse mackerel of Misaki Fishermen Union is four times higher than that of the Himeshima Fishermen Union. The main reason for such a large difference in unit price between the Unions is that the horse mackerel of Misaki Fishermen Union get the brand name "Hana-Aji," which can be sold at a high price; the other Fishermen Unions cannot use the name Hana-Aji. People in the Misaki district call the peninsula "Hana" and "Aji" is the Japanese name for horse mackerel. The price of Hana-Aji is about 1,200 Japanese yen/kg in the market but about 2,100 Japanese yen/kg in the case of direct selling. Such efforts to give a product a brand name result in high prices for the harvest at Misaki Fishermen Union.

The foregoing example suggests that the fishery in the present day does not end with the fishing work itself but that sales strategy is also very important to provide a high income for the fishermen. Moreover, active exchange between fishermen and people in the cities is very important for a fishing village to be sustainable. For example, the experiencing of fishing activity by city people is very useful to motivate the fishing village.

2.3 Traditional Legal System Related to Satoumi

Many kinds of human activities such as fisheries, marine transportation, marine sports, and reclamation interact in the coastal sea. Therefore, some troubles may happen, and such difficulties will be solved finally in the courts. The present legal system cannot solve all kinds of problems in the coastal sea, but some traditional legal systems may be useful to solve these difficulties.

From olden times, people interacted with the coastal sea near the coast to harvest marine products and use the navigation routes. Through such activities, a traditional legal system has been established. The coastal sea was a kind of commons where people cooperatively used the resources and defined some rules for the maintenance of the commons, which is called Iriai in Japanese. It is similar to the case of Satoyama (Murota and Mimata 2004).

Iriai right is a right of common use: that is, people cannot own the land and resource by themselves but can cooperatively use the land and resource, and it is different from co-owning. When people live there, they have the right of Iriai but they lose the Iriai right when they move somewhere else. Iriai right is a local wisdom of our old people to conserve the resources in Satoyma and the coastal sea (Kumamoto 2000).

The traditional legal systems in the coastal seas of Japan and Southeast Asia are introduced, and their relationship to the creation of Satoumi are discussed, in this chapter.

2.3.1 Environmental Right (Date City in Hokkaiso and Nakatsu City in Oita Prefecture)

Environmental right is "The right by which people have the good environment and can manage it," and this was established at the Tokyo International Conference on public damage in March 1970. It was declared as the right of the people to protect the environment from destruction at the 13th symposium on the people's rights in September 1970.

It was thought that legal right first arises when people observe public damage to the environment, but the new point of view says that the environmental right always exists for everyone to conserve their environment, where the people live, when some phenomena will damage the environment. The lawyer who insists on the environmental right thought that the environmental right must be established as a personal right to prevent the occurrence of public damage and to preserve the environment.

The environmental right was implemented at the trial of Date Power Plant in 1972 and the trial of Buzen Power Plant in 1973, but both resulted in failure; that is, the environmental right was not admitted.

2.3.2 *Irihama Right (Takasago City in Hyogo Prefecture)*

Mr. H. Takasaki, who was a pastor in Takasago City, thought that coastal pollution such as polychlorinated biphenyl (PCB) pollution had occurred when residents near the coast could not approach the beach after the 1960s, when the coast was occupied by reclaimed industries, and he claimed the declaration of Irihama right based on the idea that industry should not pollute the coastal sea where the people could easily access the beach, and conducted the Irihama movement (Takasaki and Takakuwa 1976):

"The sea exists for all the people from the old time and people have the right under which they enjoy walking, viewing, fishing, swimming, beach combing, collecting shells, taking seaweed, and so on at the beach. Such right exists under the social legal system. Iriai right also exists in the beach forest for wind blocking. We call all such rights "Irihama right," and we think that the Irihama Right must not be invaded because it is a part of the people's right by which they can live in good environment in the constitution.

However, the beach has been reclaimed for big industries under the government policy of economic development and the Irihama Right has been invaded, which has resulted in the occurrence of many kinds of public damage.

To prevent the occurrence of public damage, to conserve our environment, and to recover nature, we declare the Irihama Right as a part of our movement.

21 February 1975 by Hiroshi Takasaki"

The Isokusa Right, which is discussed in the next section, is a kind of Irihama Right but there is no example of a winning trial based on the Irihama Right; that is, the Irihama movement is stalled.

For example, in the case of the trial at Nagahama in Ehime Prefecture, where people insisted that fishing harbor construction be stopped to preserve the swimming beach on the basis of Irihama Right, the Matsuyama Regional Court did not admit the people to join the court in May 1978. The reason of this rejection is "swimming at a beach is not a right for the people but it is permitted because the nation does not prohibit the swimming there."

On the other hand, at Arai in Takasago, where the Irihama Right was born, a project for public access to the beach was conducted: a road 20 m wide and 800 m long, and a park 40 m wide, 300 m long, and with an area of 2.7 ha, were arranged by donations of the surrounding industries, Hyogo Prefecture, and Takasago City. This park was named "Wind park at Arai beach" and was opened in November 2006. More than 20 meetings of the association for the management of this park, consisting of local people and industries, occurred before the construction of this park (Takatani 2007). This park may be one of the results of the Irihama movement.

Our imagination is limited, and it is very important for us to directly touch the sand or water at the beach to be familiar with the costal sea.

2.3.3 *Isokusa Right (Onyu Island in Oita Prefecture)*

People at Onyu Island (population of 1,200; circumference of 22 km) in Saeki City, Ooita Prefecture, have taken seaweed (Hijiki in Japanese), sea algae (Wakame in Japanese), turbot, snails (Nina in Japanese), and so on at the shore of the Ishima-ura coast for a long time. They conducted a court procedure on the basis of Isokusa Right, which means that the local people have a right to freely take seaweed and turbot at the local beach (Nagao 2005).

Isokusa Right is considered to be a kind of Iriai (co-use) right. At Ishima-ura coast, people can take not only seaweed and turbot at the shore but also valuable sea algae such as Tengusa (in Japanese), abalone, sea cucumber, and so on in the offshore areas. Therefore, the right to take such valuable marine biota has been under free competition since about 1963. The winner can take such valuable marine biota using a set net and by other means. The money obtained from the competition has been used for the common management of the village. It reached 1,600,000 Japanese yen in older days but it has decreased to 40,000–400,000 Japanese yen in recent years.

A new development plan for Saeki Harbor was established in August 1993 where the reclamation of 6.1 ha at the Ishima-ura coast was planned for a processing area for wastes. Most of the people in the Ishima-ura area objected to this reclamation plan; that is, the vote for agreement was 93 and the vote for objection was 238 in 1998.

The reclamation was admitted legally in January 2003, and the reclamation work was begun in November 2003. The environmental assessment before the reclamation stated that there were no valuable biota in the area to be reclaimed, but Mr. H. Yamashita, who is a specialist of shell (mollusk) species, discovered 6 Red Book species and 20 new species in Ooita Prefecture at the Ishima-ura coast. Ooita Prefecture then carried out additional observations on 14–17 November 2003, but concluded in December 2003 that the results of the assessment were not changed. Ooita Prefecture tried to begin the reclamation work on 24 January 2004 because the fiscal year end of March 2004 was approaching. However, local people stopped this beginning of the reclamation work by direct action to enter into the process ship.

People in Ishima-ura area went to the Ooita regional court to ask for denial of the reclamation license in April 2003 and to ask for stopping the reclamation work in November 2003. However, the Ooita regional court gave the decision of "The Isokusa Right is not matured to be the social right" and rejected the people's demand in April 2004. People immediately applied again to the court.

Prof. K. Kumamoto of Meiji-Gakuin University, who was a surety on the side of the accusers, said at the court in February 2006 that "Custom was only custom at the beginning but it has changed to right when it matured. Isokusa Right was described in 'the investigation on fishing custom' published in 1902 and was continued in the local society after that time. Therefore it already matured and changed from custom to the right."

Moreover, he said that "The value of Isokusa Right is only a hundreds thousands Japanese yen a few years ago but it corresponds to a few billion Japanese yen stock at the present time of high benefit to the natural stock and it is the big fortune of the people at Ishima-ura area. Therefore when the Ooita Prefecture invades their fortune by reclamation, Ooita Prefecture needs the agreement of local people and must guarantee to them. Right is not given by the government but the continuous activity by the local people can confirm the right."

2.3.4 Co-use of Environment Right

Nakayama (2006) claimed that "Environmental Right" is not the passive right of being co-used or owned but the active right, being the "Right for each person to use some environment by the way of co-using and in the way of co-owning." He said that such a right must be called the "Co-using right of environment." He showed examples of such rights to be the right to catch fish and bivalves, to swim at the beach, to watch the landscape, perform navigation, and so on. The concrete manner of co-using is decided by the consensus of related people. In many cases, the way has been decided at each area on the basis of the traditional rules there, and the duty is also decided at the same time.

This right is not yet admitted by the Japanese courts.

2.3.5 Sasi

In the Indonesian language, "Sasi" has the meaning of "rest" or "prohibit" and it expresses the management system of land and marine resources (Murai 1998). For example, taking the nest of swallows is permitted to certain families, and it is permitted only three times before hatching and after nesting during July to October. Moreover, the harvest of palm trees is permitted at designated areas in designated periods to prevent the price being lowered by too much harvesting.

In the case of Ronpa, which is a fish ascending the rivers similar to salmon, Sasi continues from April when the juveniles of Ronpa approach the coast until November when they go back to the sea after spawning. The opening of Sasi is declared to catch Ronpa going back to the sea. The harvest of Ronpa continues only a few days, and the caught Ronpa are preserved for food by smoking.

For sea cucumbers, the leader of the village declares "Sasi" depending on its price or resources in the longest 5-year Sasi period.

Even in the case where the local people strictly follow such Sasi, when some people living in other villages violate Sasi, the sustainability of resources will be endangered.

I have introduced some topics on the traditional legal system such as environmental right, Irihama right, Isokusa right, co-using environment right, and Sasi.

A similar traditional legal system is found in other places; for example, the new Ishigaki Airport Construction Plan, which was expressed in 1979, was withdrawn in 1989 because of rejection by local people based on the Iriai Right to harvest bivalves and seagrass at the beach where the construction was planned.

Such a traditional legal system has been maintained by local wisdom for the local people to use the rich local resources sustainably. The traditional legal system is available only under conditions where the local people continue their productive action there; that is, a legal system that is not based on the local people's productive actions is not possible.

Local people must continue their productive actions based on the traditional legal system and must clarify the necessary duties for the conservation of local resources in the local coastal seas for the creation of Satoumi.

2.4 Short History of Satoumi Concept

I first used the terminology of Satoumi in 1989 when the editorial boards of the Japanese Society of Civil Engineering and Japanese Water Environment Society asked me to write a proposal on the future study theme in the coastal sea. I claimed that the Satoyma concept must be applied to the coastal sea to use sustainably the resources in the coastal sea, and we may call such a coastal sea Satoumi (Yanagi 1998a, b).

After that proposal, I mainly conducted research on the mechanisms of the occurrences of red tides or hypoxia, not on the Satoumi.

During study in the Seto Inland Sea, Japan, I sometimes felt need for roots for my study; that is, I must have some incentive to continue my studies when I come across some difficulty in the study. In my case, it was the friendship with the local fishermen. I wanted to see the fishermen smile, who are having some difficulties, by showing them the way to solve their problems. I worked hard to find the solution by imagining the fishermen's smiles.

The reason why I think that the fishermen are important is that the coastal sea will be destroyed if the fishermen disappear; that is, the coastal sea without the fishermen, who continue to watch and conserve it, will be used in a disorderly manner by people on land who do not understand the characteristics of the coastal sea. The fishermen are the guardians of the coastal sea.

2.4.1 Introduction of Satoumi

The Association of the Seto Inland Sea Conservation obtained the budget for Environmental Education at the Beach in the Seto Inland Sea during 2005–2007 from the Nippon Life Insurance Funding Agency. I cooperated with this project as the Chairman of the Planning Committee of Research Institute for the Seto Inland Sea and participated in education activities at eight places in 3 years. I discussed with the participating students and their parents in these environmental education activities.

Many participants wanted a textbook for environmental education in the Seto Inland Sea. Therefore, I wrote the textbook for environmental education in the Seto Inland Sea with the help of some school teachers around the Seto Inland Sea. The title of this textbook is "Seto Inland Sea: Introduction to Satoumi Concept" (Yanagi 2005a), and this textbook was received by the people with good response.

Some people who read this textbook said to me "You have to write the text of Satoumi after the introduction of Satoumi," and thus I wrote the book "Satoumi" in 2006 (Yanagi 2006).

2.4.2 *Sato-Umi*

I made a presentation on Satoumi at the International Conference of Oceanography at Stockholm in 1989. The response of the audience was very bad; some oceanographers from the United States said "Are you the servant of fishermen?" One such person claimed that "The scientists must study on the natural mechanism of its variation and must not study for someone." Of course I did not agree with him and we had a severe debate, but we could not understand each other. Such a situation has not changed in the succeeding 10 years; that is, the Satoumi concept is not admitted by scientists in Western countries.

However, the response at the time of the 7th EMECS (Environmental Management of Enclosed Coastal Seas) at Coaen, France, in 2006 was completely different when I presented the concept of Satoumi. In the plnary session on the last day, the speaker in summarizing my session said that "Satoumi is a good concept for the symbiosis of human and nature in the coastal sea and I hope that Satoumi will become the global terminology such as tsunami." I felt that the situation around the coastal sea had changed from preservation to conservation, even in the Western countries.

After coming back from France, I translated my Japanese book "Satoumi" to English in a half year and asked the publisher, Kluwer, to complete the check by a native speaker of English within 3 months. At last, "Sato-Umi: A New Concept for Coastal Sea Management" was published in 2007 by TERRAPUB in Tokyo (Yanagi 2007a).

This English book attracted some foreign readers, and the first workshop on Satoumi was held during the 8th EMECS at Shanghai, China, in October 2008. The invited speakers from Korea, China, Thailand, Indonesia, the United States, France, and Japan discussed the usefulness and limitations of the Satoumi concept. The speaker from the United States said "The Satoumi concept is useful for the paradigm shift in order to solve the conflict among the people related to oyster culture in Chesapeake Bay"; the invited speaker from France said "Satoumi concept is similar to the EBM (Ecosystem-Based Management) in EU"; and the speaker from Korea said "Scientists played a very important role in the total load management of nitrogen and phosphorus in Masan Inlet, Korea." The speaker from China said "The IMTA (Integrated Multi Trophic Aquaculture), where fish, sea algae, and sea cucumber are cultured in the same area, is successful in China"; a speaker from Thailand said “The oyster culture is conducted at the pond between shrimp pond and open ocean in order to purify the polluted water in the shrimp pond and flush out the clean water to the open sea”; one from Indonesia said “Sasi is followed by the local people but the people from the other places violate Sasi”; and a speaker from Japan said “Short-necked clam is rehabilitated at the mouth of Fushino River, Yamguchi Prefecture, Japan by the successful ICZM (Integrated Coastal Zone Management) including scientists, manager, and citizens. This is a kind of Satoumi establishment.”

In the summary discussion, it was pointed out that “Scientific knowledge is not sufficient for the sustainable use of coastal sea resources but we have to include the local tradition, culture, religion, and the countermeasure to the global trade of fisheries products into the coastal zone management for the establishment of Satoumi.”

2.4.3 *Response of Government*

The environmental managers in the central government of Japan are interested in the Satoumi concept, and it has been adopted in “21st century strategy for the establishment of environmental nation (2007.6),” “The basic plan for the ocean management (2008.3),” and “National strategy for the biodiversity (2007.7).”

The budget for the establishment of Satoumi was defrayed during 2008–2010 through the Ministry of Environment. Such a trend gave some people a chance to speak out that “The Satoumi concept is dangerous because it gives another chance of destruction of coastal environment by doing many kinds of civil engineering using the governmental budget under the concept of Satoumi.” Such people say that “Biodiversity largely decreased mainly due to many kinds of civil engineering works such as wall construction, reclamation, and we have to prohibit such action to increase biodiversity in the coastal sea.” We can understand that moderate human interaction may increase biodiversity in the coastal sea from the examples in Kansai Airport and the removal of starfish in the coral reefs (please refer to Sect. 2.1).

However, the population of fishermen in Japan is only about 200,000, 0.2% of all current population now, and it is decreasing. The main players in Satoumi are the fishermen who conduct productive actions in the coastal sea, but the collaboration of the other 99.8% of the people (citizens) is essential to the establishment of Satoumi. To get citizens to help, we need to consider the legal system (Yanagi 2008a), economic aspects (Yanagi 2007b), cultural aspects (Yanagi 2008b), and so on.

2.5 Style of Satoumi Study: Interaction Between Society and Science

Satoumi is defined as "the coastal sea with high biodiversity and productivity under the human interaction" (Yanagi 1998a, b), but there is no Satoumi now. The way of creating Satoumi is also not clarified. The movement of creating Satoumi will succeed by the close collaboration between scientists who propose the Satoumi concept and fishermen and citizens who will support Satoumi.

The necessary collaboration among scientists, fishermen, and citizens is discussed in this section.

2.5.1 Role of Scientists

The role of scientists is to discover something that has not been known previously and to tell it to the people in the society. Such a role is accomplished by publishing scientific papers.

Therefore, the measure for deciding the value of scientists depends on the number of high-quality scientific papers published by them. The interaction of a scientist with citizens is not evaluated as his value but indeed may sometimes be given a negative value because other scientists consider that he plays with citizens, doing nothing for science.

However, some criticism has stated that a scientific paper only discusses the tale on the table, not the fact in the society. The real study results must be reflected by the real society.

The Satoumi concept has been developed from the real coastal sea where people suffer from polluted conditions, and it aims to conserve the coastal sea by collaboration among scientists, fishermen, and citizens. The value of the Satoumi concept depends on its social success, that is, the success of the creation of Satoumi.

2.5.2 Role of Fishermen

The wisdom of fishermen is invaluable to increase productivity in the coastal sea. For example, the ways of increasing productivity in the line fishing ground and in the trawling ground are completely different. In line fishing, a clean fishing ground without spreading bait is important, but increasing the biomass of bottom fish by spreading the bait is important in the trawling ground. The necessary information in a natural fishing ground and an aquaculture ground are also different; for example, information on suitable water temperature and salinity for migrating fish are useful in the natural fishing ground, but information on the frequency of occurrence of a swift current that may sweep away the remaining bait and fecal pellets is useful in the aquaculture ground. The wisdom of increasing productivity in the seaweed culture ground and the oyster culture ground are different. A high nutrient concentration is preferable in the seaweed culture ground, but too high a nutrient concentration is not good in the oyster culture ground because a red tide may occur as a result of eutrophic conditions and the cultured oysters will die.

Such situations suggest that the creation of Satoumi must include many kinds of wisdom from the local fishermen.

The information and actions necessary to increase biodiversity are also different in the line fishing ground and the trawling ground, the natural fishing ground and the aquaculture ground, and the seaweed culture ground and the oyster culture ground; for example, catching herbivorous fish such as Aigo in Japanese results in the conservation of seagrass or sea algae beds and high biodiversity in the line fishing ground, and using some fishing nets with a coarser mesh size results in high biodiversity by allowing the escape of small fishes in the trawling fishing ground. Designating Marine Protected Areas (MPAs) is also important because we do not thoroughly understand the ecosystems in the coastal sea. Therefore, we have to protect some part of those ecosystems without touching them.

When the fishermen try to increase productivity only for their own sakes, the coastal sea ecosystem can easily collapse. High bioproductivity will be conserved by keeping a balance between the ecosystem and biodiversity in the coastal sea.

Not only the aid from scientists but also the collaboration with citizens are important to keep the balance of the ecosystem and biodiversity in the coastal sea.

2.5.3 Role of Citizens

About 30 million people live in the watershed of the Seto Inland Sea, Japan, but the population of fishermen, who are the main players in Satoumi, comprises only 30,000, that is, 0.3% of the total population. It is impossible for 0.3% of the people to create Satoumi. Cooperation between fishermen and citizens who have no direct relationship to Satoumi is very important for the creation of Satoumi.

Citizens must be familiar with the coastal sea, understand the importance of the conservation of coastal sea for their lives and culture, and must not dump wastes into the coastal sea. Moreover, discussion and management between fishermen and citizens who enjoy sports fishing is necessary in the coastal sea where the sports fishing catch is large.

Citizens must understand the meaning of increasing biodiversity and productivity in the coastal sea and must not decrease biodiversity and productivity in the coastal sea by their actions.

It is not easy for the citizens to act in the aforementioned fashion by themselves; local authorities or local government officers must guide the fishermen and citizens to such action as exchange of information, beach cleanups, environmental education, and so on.

2.5.4 Regional Culture Power

The topic of rock fish in Kagawa Prefecture is very interesting as a success story of a good relationship between science and society.

Rock fish, called Takenokomebaru in Japanese, was one of the main local fish in Harima-nada and Bisan Strait, the central part of the Seto Inland Sea, until about 1955, but its harvest drastically dropped during 1955–1965 as a consequence of water pollution caused by rapid industrialization. The scientists in the Kagawa Prefectural Fisheries Experimental Station tried artificial egg collection, juvenile production, and local release to recover the Takenokomebaru harvest, and their trials have succeeded so that its harvest began to increase recently.

However, women in Kagawa Prefecture do not buy Takenokomebaru because they do not know how to cook the fish, meaning that the cooking of Takenokomabaru disappeared from the culture of Kagawa Prefecture. When there is no sale of Takenokomebaru, the cost of artificial breeding and juvenile production is not compensated by taxes, and the scientists cannot continue their activities (Shono 2007).

Then, the Kagawa Prefectural Fisheries Experimental Station collaborated with the Kagawa Fishermen Union and began to teach the way to cook Takenokomebaru to the women in Kagawa Prefecture, and asked them the style in which they preferred to buy the fish (raw, cut into pieces, sliced pieces, baked or fried, etc.). After such activities, women began to buy Takenokomabaru in the local market.

The recovery of the harvest of rock fish, called Takenokomebaru, means increasing biodiversity in Harima-nada and Bisan Strait; that is, the collaboration among scientists of the Kagawa Fishery Experimental Station, staff of Fishermen Union, and the supermarket results in successful increase of biodiversity in the coastal sea. Local culture, including cooking habits, is very important to increase biodiversity in the coastal sea.

This topic suggests that a narrow point of view is not useful to create Satoumi because Satoumi creation relates to the total aspects of society, and we must integrate all activities by the people related to the coastal sea to create Satoumi.

2.6 Creation of Satoumi as Rehabilitation of the Coastal Sea

Many shallow areas in Japan were reclaimed after the execution of the "New Industrial City Law" and "Establishment of Special Area for Industries Law" in the 1960s, and deposits of wastes from coastal lands resulted in heavy water pollution in the Japanese coastal seas. For example, grey mullet with a curved backbone swimming at Mizushima-nada, and gobies with cancer, were caught at Kure Bay in the Seto Inland Sea at that time.

The central government enacted the "Prevention of Water Pollution Law" in 1970, and the disposal of wastes into the coastal sea was regulated. As a result, the Japanese coastal seas could escape from its "Dead Sea" state.

Opinions on conservation of the global environment became mainstream after the Earth Summit in Rio de Janeiro, Barajil (Brazil), in 1992. The Japanese government enacted the "Basic Environmental Law" in 1993 and executed the "Environmental Assessment Law" in 1997 to realize sustainable development. However, environmental recovery was not remarkable, and thus the Japanese government enacted the "Promotion of Nature Recovery Law" in 2002 and encouraged cooperation among local peoples and nongovernmental organizations (NGOs), the development of adaptive management.

Special committees for the recovery of a natural environment were established in Tokyo Bay, Ise Bay, Osaka Bay, and Hiroshima Bay, based on the "Promotion of Nature Recovery Law," and several actions have been ongoing, but the situation is not good, as mentioned in the following examples.

2.6.1 Rehabilitation of Osaka Bay

A special committee for the rehabilitation of Osaka Bay was established in 2003, and the Ministry of Transportation and Infrastructure, Ministry of Forest and Agriculture, Ministry of Industry and Economics, Ministry of Environment, Shiga Prefecture, Kyoto Prefecture, Osaka Prefecture, Hyogo Prefecture, Nara Prefecture, Wakayama Prefecture, Kyoto City, Osaka City, Kobe City, and the Development of Osaka Bay Area Organization constitute this committee. Management was conducted by the Kinki Regional Office of Ministry of Transportation and Infrastructure. The aim of this committee is to create a plausible Osaka Bay by recovering a friendly and rich coastal sea through a good network from mountains to rivers to the coastal sea.

The concrete aim and indicators are (1) maintaining dissolved oxygen (DO) at the bottom layer at more than 5 mg/l throughout the year for the survival of the benthos, (2) expanding the areas of tidal flats, seagrass beds, and shallow areas for the living space of marine biota, (3) maintaining chemical oxygen demand (COD) at swimming beaches at less than 2 mg/l for familiar nearshore areas, (4) increasing natural beaches for the citizens to become familiar with the coastal sea, (5) expanding the green areas adjacent to the beach as recreation places for city people, and (6) creating beaches without garbage.

To realize these aims, (1) reduction of the land-based load of pollutants, (2) management of marine environmental improvement, and (3) monitoring of the recovered Osaka Bay were considered.

However, red tides in the upper water layers and hypoxia in the bottom layer occur in summer even now. Moreover, there are few beaches where people can have access (Fig. 2.8). Garbage is present on the beaches of Osaka Bay throughout the year (Yanagi 2009a).

The situations in Tokyo Bay, Ise Bay, and Hiroshima Bay are similar to that of Osaka Bay.

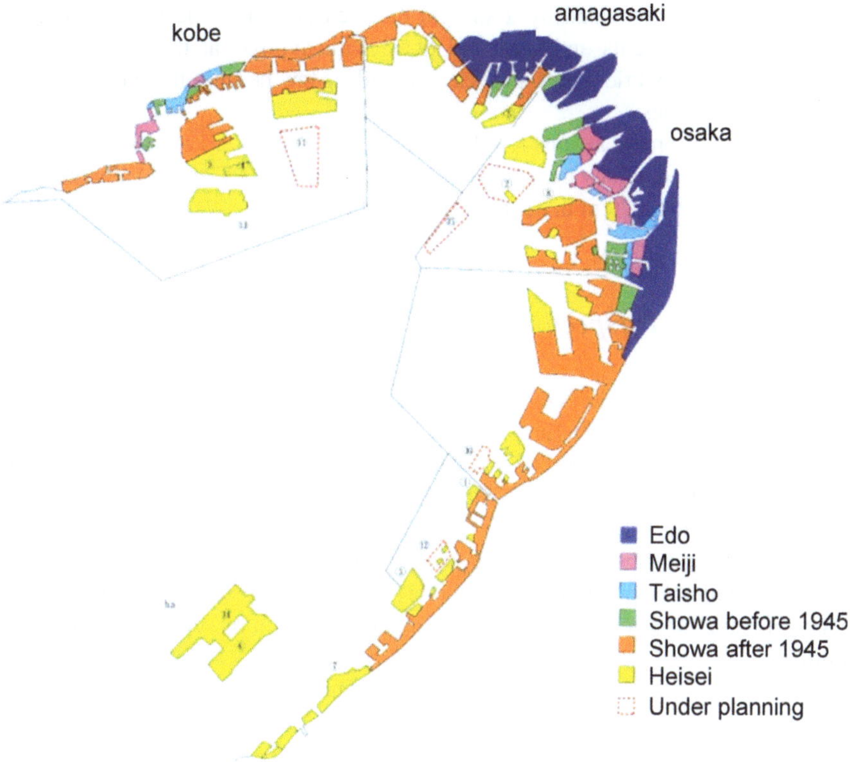

Fig. 2.8 Reclamation in Osaka Bay (Preservation Association of the Seto Inland Sea 2007)

2.6.2 How Do We Rehabilitate the Coastal Sea Environment?

We, as human beings, live on land and do not know the sea well. Without correct understanding about the sea, we reclaim the coastal sea, release wastes and garbage to the coastal sea, and use the coastal sea as we want. Those actions have resulted in the deterioration of the coastal sea. Yanagi (1998a, b, 2007a) claimed that the Stoumi, which is a coastal sea with high biodiversity and bioproductivity under human control, must be established as the target of rehabilitation of the coastal sea.

We have to understand that dense, long, and orderly cycling of compounds in the coastal sea is necessary to establish Satoumi. We have to realize the many kinds of habitats in the coastal sea by creating appropriate spaces for the living marine biota and stopping the succession of habitats to the climax condition (see Sect. 2.1).

Shallower areas of less than 20 m in depth, where sunlight can penetrate to the sea bottom, play an important role for orderly material cycling because the attached diatoms grow and assimilate the nutrients there, becoming the base for the benthos.

In the coastal sea bounded by a straight elevated coastline, the floating diatoms assimilate the nutrients and form red tides, with hypoxia at the bottom layers. Productivity without tidal flats and seagrass beds is low because there is no nursery ground for juveniles.

We have to promote actions for (1) establishing many routes of material cycling, (2) increasing biodiversity by increasing the different habitats, and (3) rehabilitating areas shallower than 20 m in depth to increase biodiversity.

2.6.3 Future Perspective

The Satoumi concept has been accepted by many people, and it is included in the "Strategy of 21st Century Environmental Country" (2007.6), "The 3rd National Strategy for Biodiversity" (2007.11), "Basic Marine Plan" (2008.3), and "Strategy for Economic Development" (2008.6). Some budgets were allocated to the activities of the creation of Satoumi.

In foreign countries, the satoumi concept (Yanagi 2008b) is accepted in a favorite sense and with the opinion that people must conserve the coastal sea by the coexistence of nature and humans instead of trying to preserve it by separating nature and humans.

It is expected that discussions on environmental conservation in the coastal sea will arise.

2.7 Co-owning and Deepening of the Satoumi Concept

The symposium "Co-owning and Deepening the Concept of Satoumi in Japan" was held at the Research Institute for Applied Mechanics, Kyushu University, on 9 October, 2009.

The word "Satoumi" was used first by Yanagi (1998a, b), but it has been used independently by many people or groups that are related to the conservation activities of coastal seas in Japan after 2000. The meaning of Satoumi as used by different people or groups is different, and a common opinion on the definition and necessary action for satoumi creation has not been obtained.

Under such conditions, people or groups that use the word satoumi gather at one place for the discussion for co-owning and deepening the satoumi concept.

The content of the symposium is introduced in Table 2.3.

Table 2.3 Program of the symposium "Co-owning and Deepening of the Satoumi Concept in Japan"

Symposium at the Research Institute for Applied Mechanics, Kyushu University "Co-owning and Deepening of the Satoumi Concept"	
Chair: S. Kakuma (Okinawa Prefecture); Co-Chair: T. Yanagi (Kyushu University)	
9:30–18:00 on Oct. 9, 2009 Room W601 at RIAM, Kyushu University	
Moderator: T. Yanagi (Kyushu University)	
9:30–10:00	"Introduction and Satoumi in Okinawa" S. Kakuma (Okinawa Prefecture)
10:00–10:30	"Satoumi Rehabilitation by Local Community of Shiraho" M. Kamimura (WWF Coral Reef Center)
M. Kamimura (WWF Coral Reef Center)	
10:30–1:00	"Satoumi and Satohama Movement at Nakatsu" Y. Ashikaga (Playing at Beach (NPO))
(Playing at Beach (NPO))	
11:00–11:30	"Creation of Marine Forest" M. Kanda (Kuroshio Feeling Center (NPO))
(Kuroshio Experience Center (NPO))	
11:30–12:00	"Friend of Fishermen Activities" Y. Mastuda (Oshima Fishermen Union)
(Oshima Fishermen Union)	
12:00–13:00	Lunch
Moderator: O. Matsuda (Research Institute of Seto Inland Sea)	
13:00–13:30	"Rehabilitation of Tokyo Bay by Fishermen and Citizens" T. Kinman (Rebuilding of Utase-Bune in Tokyo Bay)
13:30–14:00	"Preservation of Local Fisheries Resources by Shiriya Fishermen Union" M. Inui (Suidosha Corp.)
14:00–14:30	"Satoumi and Local Rule" M. Nakajima (Mana Press)
14:30–15:00	"Management of Satoumi from the Viewpoint of Plant Ecology" S. Arai (Research Institute of Sea Weed and Algae)
15:00–15:30	Coffee Break
Moderator: S. Kakuma (Okinawa Prefecture)	
15:30–16:00	"Satoumi and Moku (Sea weed and algae)" T. In-nami (Aichi University)
16:00–16:30	"Small Fish and Local Economy" G. Setoyama (Document writer)
16:30–17:00	"Sub-global Assessment of Satoyama and Satoumi" O. Matsuda (Research Institute of Seto Inland Sea)
17:00–17:30	"Bio-diversity and Human Interaction" T. Yanagi (Kyushu University)
17:30–18:00	General Discussion (Chairs: S. Kakuma and T. Yanagi)
18:00–20:00	Party

2.7.1 *Various Satoumis*

2.7.1.1 Satoumi in Okinawa

Dr. S. Kakuma (Okinawa Prefecture) introduced the aim of this symposium and the activities in Okinawa to manage the coastal sea by using the satoumi concept. His definition of Satoumi is "Coastal sea with the good relation between human and coral reef." This definition is based on the history where people use the area between the breaking-wave zone of the coral reef and the coast as a commons. However, troubles have occurred between the local people and fishermen because of the harvest of shellfish and seagrass in this area. Under fishing law, the fishing rights to shellfish and seagrass in this area belongs to fishermen, but local people historically obtained shellfish and seagrass in this area; that is, this conflict has resulted from disagreement between the modern legal system and historical custom.

Dr. S. Kakuma claimed that he wants to recover the fishing resources in this area by applying the concept of Satoumi because the fish resources in this area have decreased.

2.7.1.2 Satoumi at Shiraho

Mr. M. Kamimura (WWF Coral Reef Conservation Center) introduced the center that established the "Committee for the Conservation of Coral Sea in Shiraho Village" in 2005 to conserve the coastal sea between the surf zone and the coast and to increase the fisheries resource there, and reconstructed the Ishihimi (Nagaki in the Okinawan language), which is a kind of "fish trap" (Kamimura 2007). The committee (1) planted a kind of ginger tree around the fields to prevent the flow of red clay, which damages the coral reef, (2) released juvenile shellfish and set up a MPA (Marine Protected Area), and (3) established the local rule that is followed by the local people and the tourists for the conservation of the environment of Shiraho area. His definition of Satoumi is "Relation between people and coastal sea where the people who have the wisdom of sustainable resource use and can receive the good ecosystem service from the coastal sea."

2.7.1.3 Satoumi and Satohama at Nakatsu

Ms. Y. Ashikaga (NPO Group of Enjoying Beach) said that the Nakatsu tidal flat at the southern part of Suo-Nada in the western Seto Inland Sea was a familiar place for excursions in spring and autumn, for getting food for dinner, and for getting fuel to heat the baths for the local people. However, local people do not approach the Nakatsu tidal flat because they forgot the old familiar place, dumped garbage there, and and the tidal flat has been changed to a place where people must not approach a place to be reclaimed because it is useless. The NPO group is conducting field observations, biota investigations, beach cleanup, fishing experiences, and so on to shorten the distance between the local people and their tidal flat.

The characteristic activity is the revival of the bamboo fish trap (Sasahibi in Japanese), which is a kind of Ishihimi made from bamboo with a length of about a few meters that existed 50 years ago. However, Sasahibi was abandoned because the seaweed net became very popular at Nakatsu tidal flat in recent years. Many kinds of biota such as shrimp, crab, and short-necked clam gather around bamboo fish traps (Sasahibi), and the local children received money by getting and selling such biota. Sasahibi increases not only productivity but also biodiversity. People near the coast go to the mountain areas to cut the bamboo for Sasahibi, and a good network between the beach people and mountain people was established by building Sasahibi. Cutting bamboo is also useful to maintain good Satoyama. The group succeeded in the revival of Sasahibi in 2008 by the help of old fishermen and the local government. The NPO investigated the effects on the reproduction of clam juveniles, the effect of increasing biodiversity, and the environmental education effects of Sasahibi construction from 2008 to 2012.

The definition of Satoumi by Ms. Y. Ashikaga is "The coastal sea with fine biota, children and fishermen."

2.7.1.4 Satoumi at Kashiwajima Island in Kochi Prefecture

Dr. M. Kanda (NPO Kuroshio Feeling Center) defines the Satoumi as "The coastal sea where people not only receive its ecosystem service but also culture, take care and reserve the resources" and introduced several activities at Kashiwajima Island in Ohtuki town, Kochi Prefecture, which he believes to be a museum as a whole. Among them, the project to construct an artificial reef for squid spawning is very interesting.

Aori-ika, in Japanese, a kind of squid, come to the coastal sea around Kashiwajima Island for spawning in early summer, but its numbers have drastically decreased in recent years because of the decreased areas of sea algae beds. Therefore, the people, divers, fishermen, wood cutters, children, local government, and NPO in Kashiwajima Island decided to construct an artificial reef for squid spawning. Wood cutters cut many branches from the mountain forests and divers fixed them to iron rods on the sea bed, where a maximum of 15,000 squid eggs were spawned. The marine biologist Dr. M. Kanda decided the most suitable place for the artificial reef based on his studies. Children could feel a connection between mountain and sea by watching the success of this artificial reef for squid spawning.

2.7.1.5 Satoumi at Mikuni in Fukui Prefecture

Mr. Y. Matsuda (Oshima Fishermen Union) reported the change of Satoumi in Mikuni City that results from the aging of the Ama (diving fishery women). Amas in Oshima Fishermen Union not only harvested seaweed on the rocks in spring, sea algae (Wakame in Japanese) in early summer, and sea urchins in midsummer, but also worked polishing the surface of the rocks for seaweed cultivation and cutting seagrass for sea algae (Wakame) culture. However, such hard work became impossible because of the aging of the Ama and changes in the ecosystem, and the productivity decreased; thus, Satoumi at Mikuni is now in danger.

Mr. Matsuda proposes to rehabilitate Satoumi by cooperation among the Fishermen Union including the Ama and the local community including children. He tries to open a part of the fishing ground and includes the citizens in the fishing activities. His definition of Satoumi is "The coastal sea where people receive its ecosystem services and conserve resources by the cooperation of fishermen and citizens."

2.7.1.6 Satoumi in Tokyo Bay

Mr. T. Kinman (Committee for Revival of Classic Fishing Boat in Tokyo Bay) stated the reasons for the revival of the classic fishing boat in Tokyo Bay, that is, (1) to continue the technology of construction of fishing boats in Japanese style, (2) to conserve wood resources, (3) to rehabilitate the shrimp fishing ground in Tokyo Bay, and (4) to conduct environmental education by inviting children to see this fishing boat.

The committee was established in 2004 and has continued several kinds of activities such as (1) seaweed production, (2) observing the tidal flats, (3) rehabilitating the short-necked clam and sillaginoid resources, (4) eco-touring from the river mouth to the mountain, and so on.

His definition of Satoumi is"the coastal sea where people can feel the relation between their life and the sea, the fishing resource for the fishermen's life is conserved there and citizens can enjoy the biota, plants and sea-scape there."

2.7.1.7 Satoumi at Shiriya in Aomori Prefecture

Mr. M. Inui (Suido-sha) introduced activities for more than 100 years at Shiriya Fishermen Union in Aomori Prefecture, that is, (1) construction of forest on the deteriorated coastal area, (2) rehabilitation of sea algae (Konbu in Japanese) beds destroyed by the eruption of the Komagatake volcano by setting an artificial reef, (3) conservation of sea algae (Konbu) beds by prohibiting its direct harvest except by catching sea algae (Konbu) that has drifted to the beach, (4) continuous monitoring of the fishing ground, and (5) control of grazing pressure of sea urchins on sea algae (Konbu) by moving the sea urchins. He also pointed out that such activities are based on the local philosophy of Sanyo-Kai in Japanese (local organization by young fishermen from ages 16–42 years), which is "Please study during the time of not-working."

His definition of Satoumi is "The coastal sea where people use the bio-resources based on the local rule and such activities result in high productivity and biodiversity."

2.7.1.8 Satoumi at Oki Island

Mr. S. Arai (Research Institute for Sea-Algae) defined Satoumi as "Sea-algae beds under the transition state by human interaction" in a narrow sense and "Landscape of fishing village and fishing ground based on the economic activities in sea-algae beds" in a broad sense.

He reported that (1) people cut a part of the sea algae (Nokogirimoku in Japanese) beds before the climax stage and cultured different kind of sea algae (Wakame in Japanese) at the gaps 50 years ago in the district of Oki and Izumo, Shimane Prefecture; (2) another sea algae (Arame in Japanese) was cut for food, and many juveniles gathered in the gaps and at the rim of sea algae beds; and (3) young sea algae grow quickly there and sea urchin and abalone can also survive is these spaces.

The management technology of sea algae beds was transmitted in 1950 to this district but it has been lost now. On the other hand, new technology based on the scientific knowledge of plant ecology has spread in recent years; for example, a kind of sea algae (Akamoku in Japanese) bed was formed by planting young Akamoku on an artificial rock reef, which is protected from grazing by fish by planting fish eating sea algae that mainly graze the old algae at the Oki district.

2.7.1.9 Moku and Pine Tree

Seagrass and sea algae caught from the seabed are called "Moku" in western Japan, and Prof. T. In-nami (Aichi University) pointed out that Moku in the sea has a similar character as pine trees on land.

People prohibit the cutting of pine trees, continue to cut the grasses under the pine trees, prevent the climax stage of the pine tree ecosystem, and define the opening day of collecting the pine tree reef to conserve the pine tree resources. A landscape with pine trees is a symbol of the symbiosis of human and nature.

His definition of "Satoumi" is "Preserved coastal sea by the people for their survival with nature and culture."

2.7.1.10 Satoumi at Iwaki in Fukushima Prefecture

Mr. G. Setoyma (documentarist) reported that fish-processing industries in Iwaki City, Fukushima Prefecture, buy cheap fish, dry them during the fresh state by employing local old people, and thus maintain the local Satoumi.

His definition of Satoumi is "the coastal sea used sustainably in order to support the life of local people." He also pointed out the importance of considering the coexistence of human and nature by negotiating with each other.

2.7.2 Co-owning and Deepening

Satoumi introduced in the former section have different definitions and characteristics but they are not contradictory to each other.

Prof. O. Matsuda (Hiroshima University) said that "The only definition of Satoumi is not necessary and different definitions of satoumi are reasonable due to different relations between people and nature at different places," and Mr. T. Kinman (Committee for revival of local fishing boat in Tokyo Bay) said that "The saroumi at each place must be defined based on the agreement of local people because the histories of the coastal sea management at each place are different."

Moreover, Mr. M. Inui (Suidosha) said that "It is important to propagate the satoumi concept in order to conserve the coastal sea because the environmental condition of coastal sea has been deteriorated due to the decrease of use by the local people."

This symposium was the first one where people related to satoumi activities gathered, and it was very successful because all the attendants presented their activities, intense discussion was conducted, and the same conclusion was reached.

At the same time, some future study themes were clarified by this symposium, such as (1) not only natural science but also social and cultural sciences must be included for the creation of satoumi because we have to clarify the desirable relationship between local people and local nature, and (2) it is also useful to establish artificial tidal flats in urban navigation channels and continue monitoring the changes of living biota there for the environmental education of city children.

I hope to convene a second symposium in the near future for deeper discussion.

2.8 Satoumi Concept as Natural Science

It is not necessary to establish one definition of Satoumi in Japan or in the world, as pointed out by Yanagi (2010), but there is one necessary keyword for defining Satoumi from the aspect of natural science, and that word is "biodiversity." High biodiversity results in high bioproductivity, and fishermen, who monitor the coastal sea environment and conserve it, can harvest the remaining marine biota after their recruitment.

It is already clarified by Yanagi (2009b) that appropriate human interaction can increase biodiversity in the coastal sea, that is, high biodiversity can be maintained by many kinds of habitats which people can manage. Chikuiso (setting large rocks on the shallow coast in Japanese) and artificial reefs correspond to managing habitats in the coastal sea. Moreover, interaction to stop the progression to climax stage in the seagrass beds increases biodiversity in the coastal sea and is similar to human interaction to nature at Satoyama. At the eelgrass beds in western Japan, about 50 years ago people commonly cut some of the eelgrass beds to use for fertilizer on field crops on land, and such action results in the gathering of small fish at the gaps in the eelgrass beds. However, such harvesting in eelgrass beds ceased because of the use of

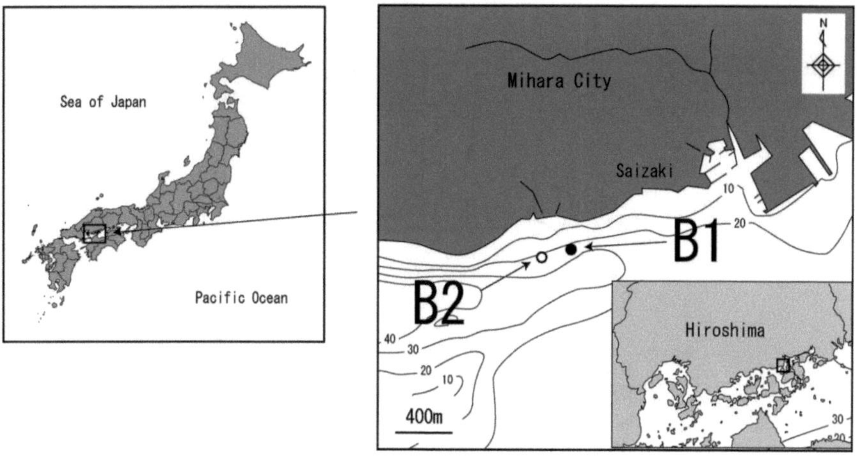

Fig. 2.9 Observation area. *B1* and *B2* refer to sampling stations

chemical fertilizers, and biodiversity in the eelgrass beds thus decreased from the loss of habitat provided by the gaps in the beds (Tanimoto, 2010).

One concrete example showing the relationship between human interaction and increasing biodiversity is introduced in the next section.

2.8.1 *Biodiversity and Human Interaction*

The artificial reef made from shells that were placed at an old sand mining area is selected as an example of the relationship between human interaction and the coastal sea and biodiversity. We examined the seasonal variation in the benthic community at the place of the shell reef setting and at a distance of 200 m (Fig. 2.9). Here I introduce the results of field observations conducted in 2010.

At Sta. B1, where seven shell reefs made of polyethylene pipe containing shells (Fig. 2.10) were set, and at Sta. B2, 200 m away from Sta. B1, we set a seine net for fish catching and sampled the mud to investigate the benthos. The reason why Sta. B2 is at a distance of 200 m from Sta. B1 is that the effect of an artificial reef does not extend to 200 m from its setting site (Itoh et al. 2008).

Vertical profiles of water temperature, salinity, turbidity, and fluorescence at Sta. B1 and Sta. B2 were nearly the same, and are vertically well mixed through the year because of the strong tidal currents, which attain 1.0 m s^{-1} at flood and ebb max. We compared caught fish and benthos at both stations. The benthos was sampled three times using a cylinder sampler 20 cm in diameter, 5 cm deep, and 50 cm long; samples were sorted by a 1-mm-mesh net.

For species and individual numbers of fish caught (Fig. 2.11), and benthos species and their individual numbers and wet weights (Fig. 2.12), all results at Sta.

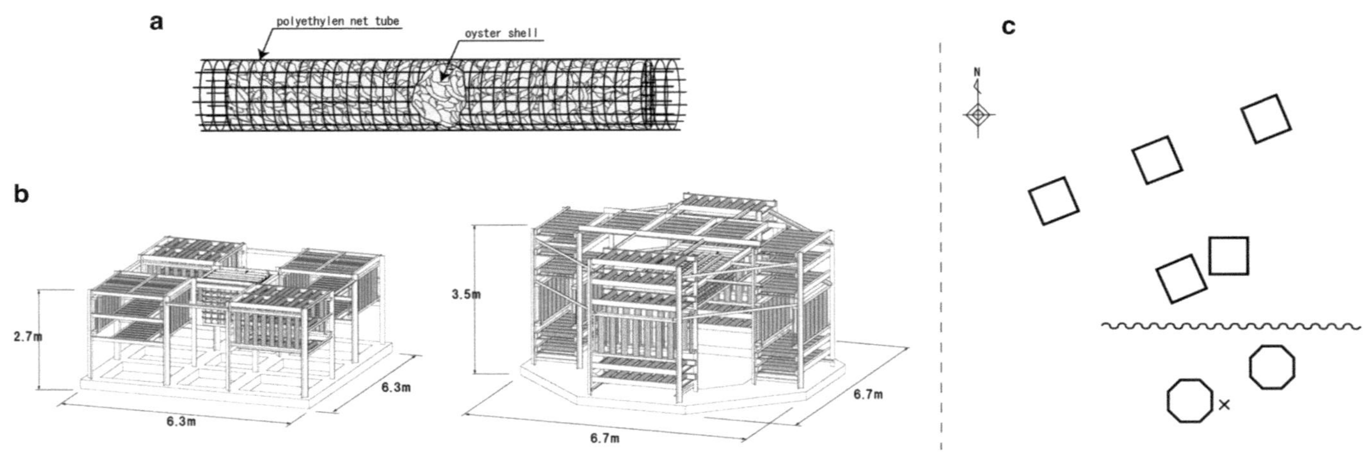

Fig. 2.10 Artificial reef made by oyster shell. (**a**) pipe and shell, (**b**) squared type reef and hexagon type reef, (**c**) arrangement of two types of reefs and x denotes Sta.B1

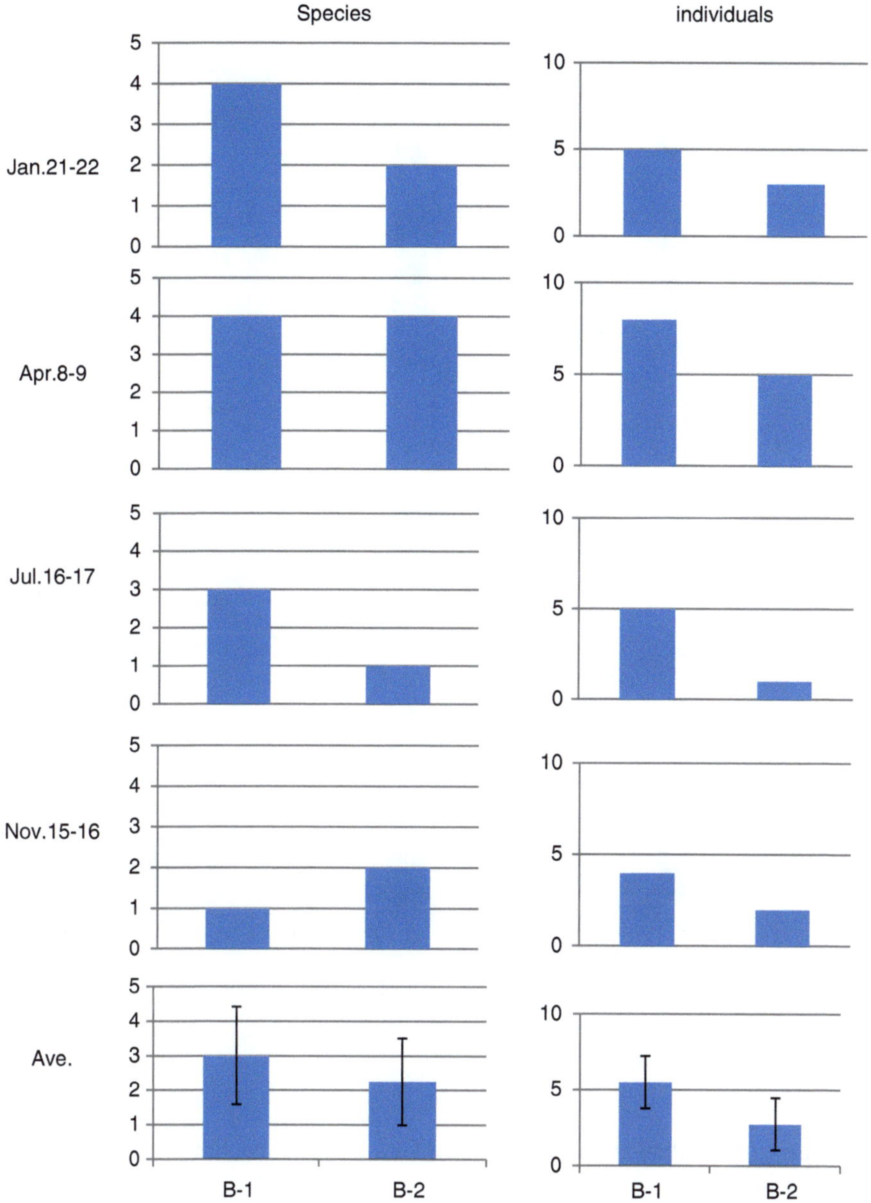

Fig. 2.11 Fish species (*left*) and individual numbers (*right*) caught by seine net at Sta. B1 and Sta. B2

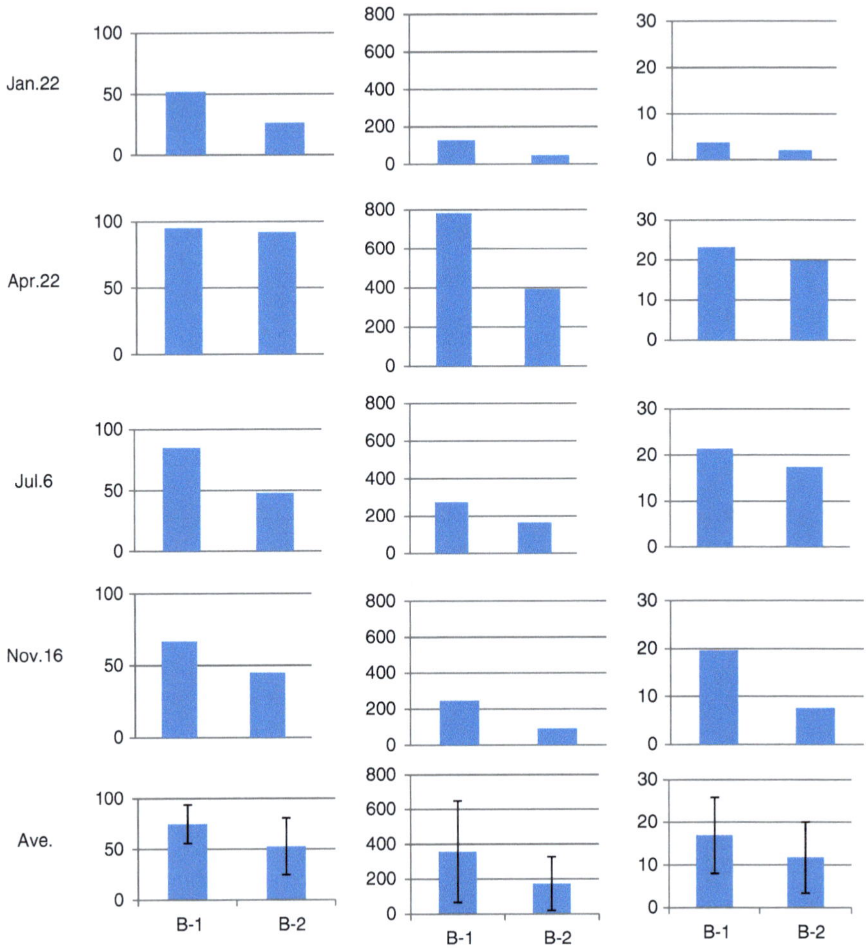

Fig. 2.12 Benthos species (*left*), individual numbers (*center*), and wet weight (g/0.3 m^2, *right*) at Sta. B1 and Sta. B2

B1 are greater than those at Sta. B2. This observation indicates that proper human interaction (artificial reef setting in this case) increases biodiversity and production in the coastal sea. The seasonal variations in benthos biodiversity index (the Shannon–Weaver biodiversity index, H'):

$$H' = -\sum p_i \ln p_i \tag{1}$$

$p_i = n_i/N$ (n_i = individual number of species I, N = total number of individuals)

at both stations are shown in Fig. 2.13. Yearly averaged index at St. B1 is larger than that at Sta. B2, and this result is statistically significant (Kendall test, $P > 0.1$).

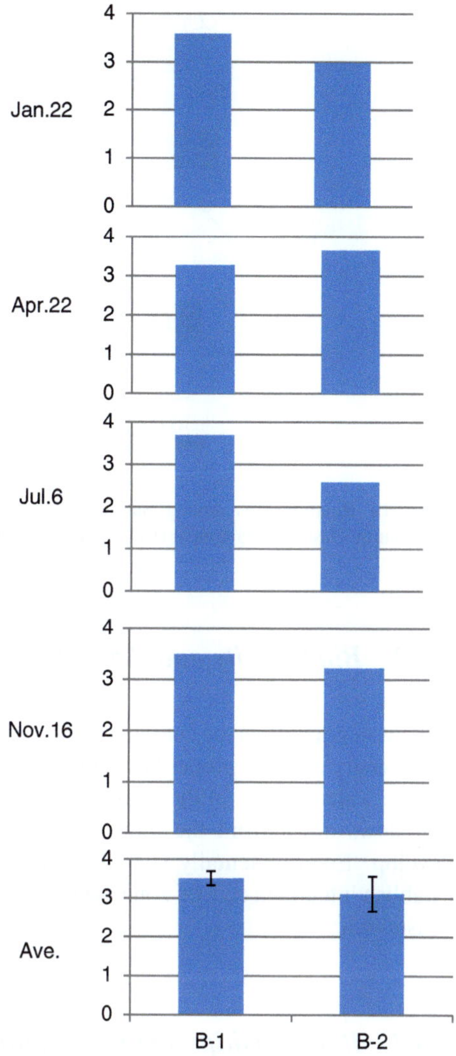

Fig. 2.13 Biodiversity index at Sta. B1 and Sta. B2

These results suggest that a new benthos ecosystem has been established after the setting of artificial reefs in 2007–2008 at the sea bottom, which was renewed by the cessation of sand mining in 1998.

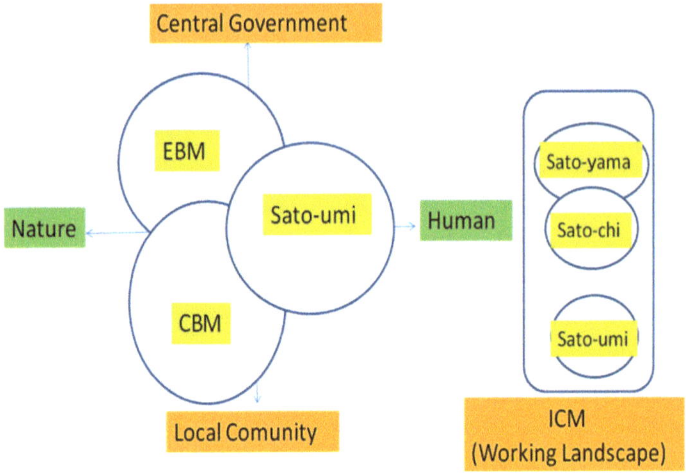

Fig. 2.14 Relationships among Satoumi (*Sato-umi*), Ecosystem-Based Management (*EBM*), Community Based Management (*CBM*), and Integrated Coastal Zone Management (*ICM*)

2.8.2 *Biodiversity and Water Quality*

If the water quality in the coastal sea is not good for marine biota, we cannot increase biodiversity there by proper human interactions.

The water quality in the coastal sea mainly depends on that of the river water that flows down from the mountain through the fields. Therefore, we have to develop “Environmentally friendly wood industry, agriculture and fisheries” and establish integrated coastal zone management to maintain good water quality in the coastal sea.

2.8.3 *Relationships Among Satoumi, EBM, and CBM*

The present main concept of coastal zone management in Europe and the United States is EBM. This concept shifts the main target of environmental monitoring from water quality, such as TP (total phosphorus) or TN (total nitrogen), to the marine biota, such as the area of seagrass beds or oyster catch, to establish a healthy coastal sea (Larkin 1996).

It is important to adopt local wisdom (indigenous wisdom) for conducting efficient EBM because the ecosystems at each place differ as a consequence of the heterogeneity of the coastal sea, and local wisdom well understands the ways to increase biodiversity at each location. Even for fishery resource management, the

EBFM (Ecosystem-Based Fishery Management) is proposed on behalf of Total Allowable Catch (TAC) (Pikitch et al. 2004).

EBM puts the basis of environmental monitoring at the ecosystem, such as seagrass beds, but the Satoumi concept puts the basis of coastal zone management on the working landscape of people near the coast and tries to establish a new human society that coexists with nature.

Community-Based Management (CBM) has become popular in Southeast Asia countries as an Integrated Coastal Zone Management (ICM) (Kakuma 2009a, b). CBM is conducted by the agreement of local communities on behalf of the control of the legal system of the central government. The Satoumi concept is similar to CBM but is a little different from CBM because Satoumi has a component related to the central government. The present situation, that the Sasi (Murai 1998) is followed by the local people in Indonesia but is violated by the people outside of the local communities (Mosse 2008), suggests that the guidance of central government is necessary for the successful coastal zone management.

Satoumi creation must be carried out by adopting local wisdom in broad land and sea areas. River management is necessary for maintaining good water quality of rivers that empty into the coastal sea. Such integrated management will be realized as an Integrated Management (IM) combining Satoyama, Satochi, and Satoumi (Fig. 2.14).

2.8.4 Sustainable Use and Exploitative Use

The aim of Satoumi is the sustainable use of fishery resources in the coastal sea. The change of sustainable use of natural resources in the developing closed society to nonsustainable use in the developed open society is seen at many places. One example is the environmental deterioration of tropical forests, where sustainable use by local people had long continued but such use was destroyed by the introduction of big foreign capital. Such changes are seen in the fishery resources in the coastal sea.

Chapter 3
Examples of Satoumi Creation

3.1 Rehabilitation of Clam Fishery Ground at Kuwana, Aichi Prefecture: Akasuka Fishermen Union

The giant clam has been a famous special product at Kuwana district in Aichi Prefecture. Kuwana is located at the mouth of large three rivers, Kiso, Nagara, and Ibi, and was prosperous as the port that connected Atsuta and Kuwana. A large tidal flat has developed near the mouths of the three large rivers and has become a good fishing ground for corbicula, short-necked clam, and giant clam.

The giant clam production at Kuwana, which was 3,000 ton/year in 1965, decreased in 1970 because of loss of the tidal flat by reclamation, decrease of river discharge, water pollution, and so on, and was only 800 kg/year in 1979.

3.1.1 Present Status of Japanese Giant Clam Fishery

The catch of giant clam in Japan in 2003 was 1,171 ton/year, which is 40% of that 10 years ago. The import of giant clam in 2003 was 18,769 ton/year, 16 times the total catch in Japan. The main import countries were North Korea and China. The main production prefecture in Japan was Ibaraki Prefecture, which provided 70% of the national catch, and Chiba, Mie, and Kumamoto Prefectures followed. The giant clam in Ibaraki and Chiba Prefectures was the beach clam in the open sea, but that in Mie and Kumamoto Prefectures was the Japanese clam in the coastal sea, and the imported clam was the China clam in the coastal sea.

T. Yanagi, *Japanese Commons in the Coastal Seas: How the Satoumi Concept Harmonizes Human Activity in Coastal Seas with High Productivity and Diversity*,
DOI 10.1007/978-4-431-54100-4_3, © Springer Japan 2013

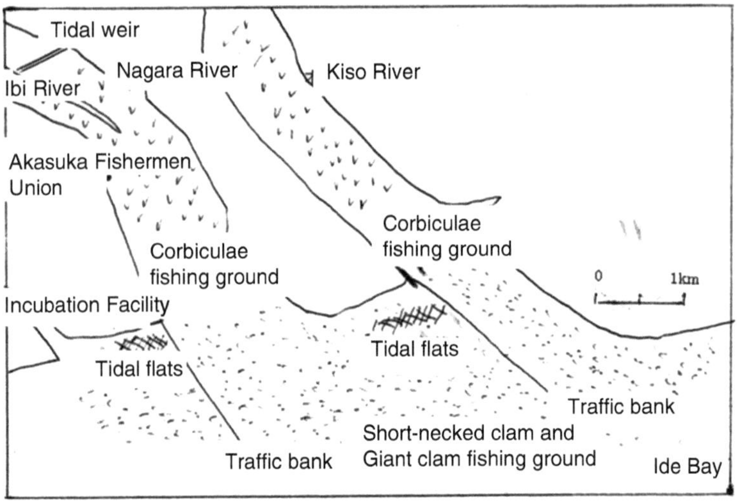

Fig. 3.1 Giant clam, short-necked clam, and corbiculae fisheries grounds and the position of the Akasuka Fishermen Union

3.1.2 Challenge of Akasuka Fishermen Union

The Akasuka Fishermen Union (the leader is Mr. Kiyone Akita) in Kuwana City, Mie Prefecture, which has a long history, 450 years, began its activities in 1990 against the decrease of giant clams to recover the clam catch.

The main fishing activities of Akasuka Fishermen Union were bivalve fisheries and fishing boat fisheries whose targets are shrimp, sea bass, sillago, flat fish, and conger eel, but the latter collapsed in about 1960 because of deterioration of the fishing ground (e.g., occurrence of oily-smelling fish). Bivalve fishing such as corbicula, short-necked clam, and giant clam has been the main fishing activity in recent years.

In 2008, 80 fishermen were engaged in corbicula fisheries, 40 fishermen in short-necked and giant clam, and 30 fishermen were engaged in seaweed culture (Fig. 3.1). The Fishermen Union restricts the working days to fewer than 3 days a week and the catch to less than 140 kg per fishing boat per day. The Fishermen Union tried artificial spawning, hatching, and breeding, and establishing giant clam aquaculture technology after 1976. They succeeded in 1987 and built a facility for breeding near the mouth of Ibi River in 1989 with funding from the Japanese Fisheries Agency that began operation in 1992 (Fig. 3.2).

In June–July of every year, giant clam juveniles were hatched out and grew in the tank, and about one million young clams about 1 cm in size were released in November on the artificial flats (20 ha × 2 sites) near the mouths of Kiso and Ibi Rivers (Fig. 3.3).

These artificial tidal flats were made in 1985 by the local fishermen who belong to Akasuka Fishermen Union under the cooperation of the Ministry of Infrastructure, Japan, using dredged sand in the estuary of the Nagara River (Tsunematsu and Kato 2008).

Fig. 3.2 Facility for giant clam juvenile incubation by the Akasuka Fishermen Union

Fig. 3.3 Artificial tidal flats constructed by the Akasuka Fishermen Union

3.1.3 Collaboration with Citizens

The staff of Akasuka Fishermen Union think that collaboration with local citizens is very important for the conservation of fishery resources such as bivalves in the coastal sea. Therefore, they have provided corbiculae to the local elementary schools for their lunch and have invited the children to the young giant clam release on the artificial tidal flats since 1975.

Moreover, they hold a fisheries festival in early July, when the taste of corbicula and clam becomes the best, and provide free miso-soup with corbicula and

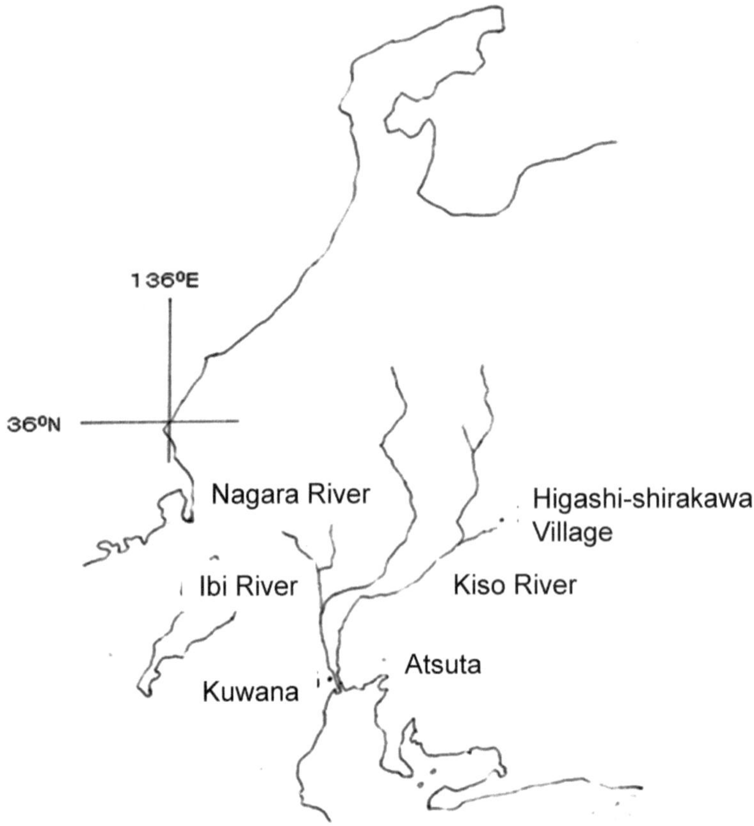

Fig. 3.4 Collaboration of forest farmers (Higashi-shirakawa) and fishermen (Kuwana)

cheap baked giant clam to the citizens for the establishment of good relationships between fishermen and citizens from 2000 on when they got a good perspective of the future of giant clam fisheries.

They think that a good forest is necessary for the conservation of good fishing grounds in the coastal sea and began cooperation with the forestry people, planting small trees on the mountain based on cooperation with fishermen, wood cutters, and the local government of Shiratori Village from 1998 and that of Shirakawa Village from 2001 (Fig. 3.4). They began to invite elementary school children in the Shirakawa village to the artificial tidal flats at Akasuka for field observation beginning in 2004.

The catch and price of giant clams of Akasuka have been stable in recent years, and the giant clam catch recovered to 90 ton/year in 2007 (Fig. 3.5). They sell all the giant clams caught through the fishermen union to prevent lowering the price. Such activities of the Akasuka Fishermen Union contribute to creating Satoumi in the estuary of the Kiso River. For detailed activities of Akasuka Fishermen Union, the reader is referred to their HP (http://www.akasuka.or.jp/).

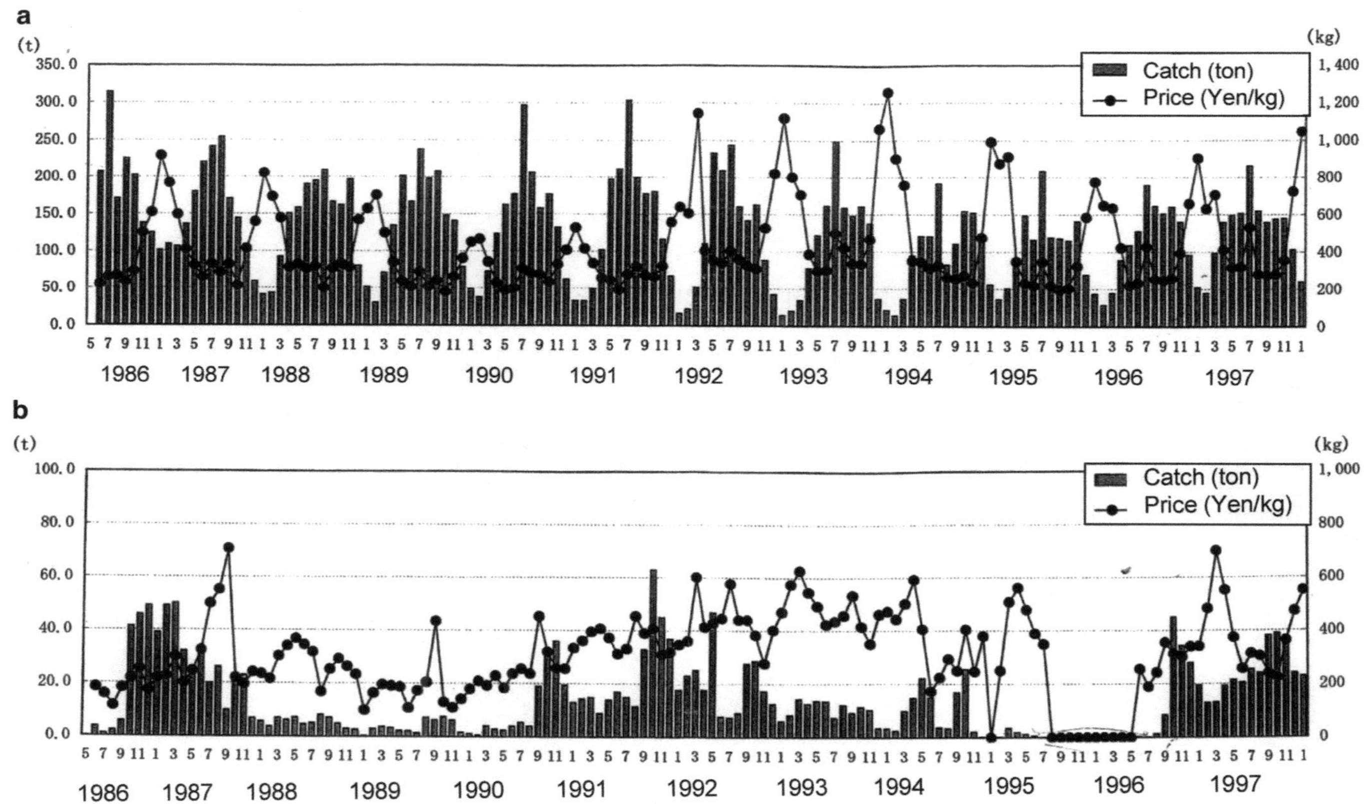

Fig. 3.5 (**a**) Temporal variation in corbicula catch and its price by the Akasuka Fishermen Union. (**b**) Temporal variation in short-necked clam catch and its price by the Akasuka Fishermen Union. (**c**) Temporal variation in giant clam catch and its price by the Akasuka Fishermen Union

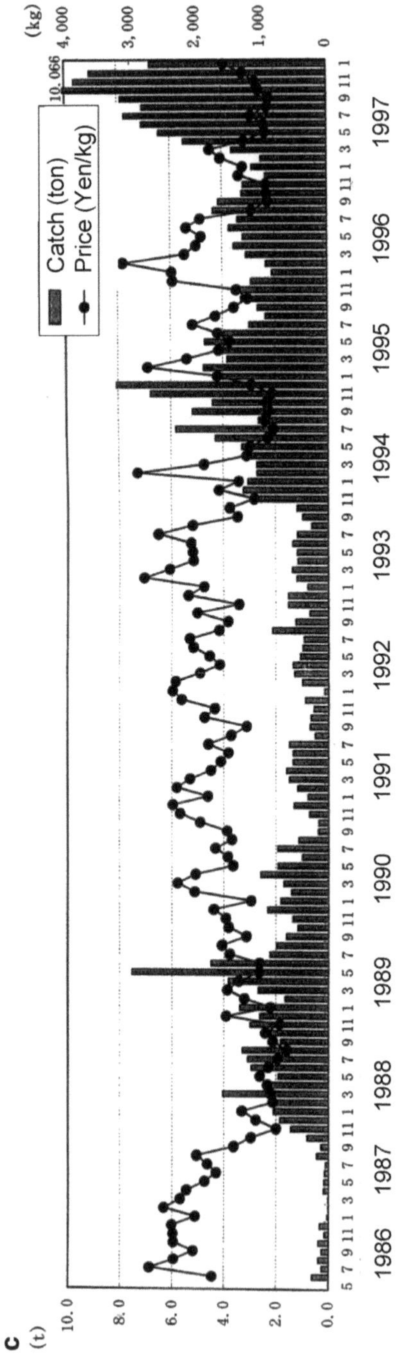

Fig. 3.5 (continued)

3.2 Rehabilitation of Eelgrass Bed and Collection of Marine Debris from the Seabed: Hinase Fishermen Union in Okayama Prefecture

The population of Hinase town, which is located in the eastern part of Okayama Prefecture, is about 11,000, and the ratio of fishermen is now only 3%, although it was once called "a big fishing town with 1,000 fishermen's houses."

The Hinase Fishermen Union numbered 107 full members and 62 submembers as of June 2008. The Hinase Fishermen Union is famous because it developed the drifting net for mackerel and the small seine net, and some fishermen moved to Nagoya in Aichi Prefecture in the east and to Korea in the west with their high technology. The seine net has been called the Hinase net. There are many relocated fishermen from Hinase in Yamaguchi, Oita, and Fukuoka Prefectures.

The Hinase Fishermen Union admits only one full member from each family, and the younger sons, except the oldest son, must migrate to other places to continue fishing. Moreover, the fishermen went to Korea and China to fish for mackerel with the drift net before the World War II.

However, the fish catch by Hinase Fishermen Union has decreased in recent years and the full members have become old. The main fishing activities of the present Hinase Fishermen Union are small seine net (about 50 families), drift net for mackerel (from April to June), drift net for bonito (July and August), small trawling net (about 50 families), oyster culture (about 50 families), and seaweed culture (2 families) (Fig. 3.6).

3.2.1 Rehabilitation of Eelgrass Bed

The area of eelgrass beds in the Hinase coastal sea area decreased after early 1960, mainly because of water pollution of the Seto Inland Sea (Fig. 3.7). The big typhoon in 1961 caused large amounts of damage to the eelgrass bed, and the eelgrass bed did not recover after that time.

Eelgrass beds form in the calm coastal sea with a sandy silt seabed. The beds can weaken the strong tidal current and strong sunlight and become the breeding place for squid and nursery grounds for small fish, because the small animals on the leaves of eelgrass are good bait for small fish.

The fishermen in Hinase Fishermen Union who were using the small seine net thought that the main reason for the decreased fish catch was the decreasing eelgrass bed area, and they began to rehabilitate the eelgrass beds under the guidance of scientists from the Okayama Fisheries Experimental Station in 1985.

Eelgrass is a plant (grass) having flowers and seed (Fig. 3.8). Eelgrass can expand its growing area by seeding and by spreading roots (rhizomes). Eelgrass seed drops on the seabed in June, germinates from November to January, and grows until July. Eelgrass ceases growth in summer and becomes a drifting grass, but it grows again in autumn (Fig. 3.9).

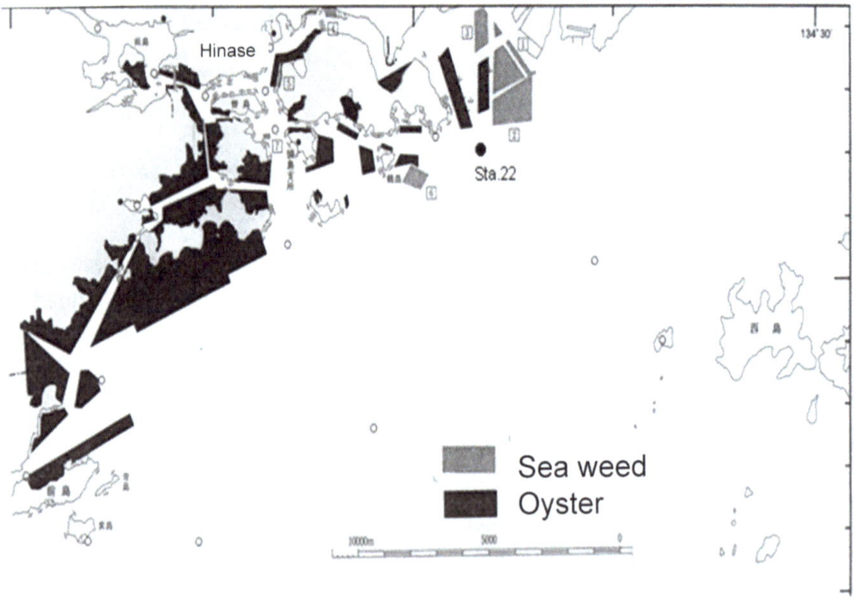

Fig. 3.6 Seaweed (*gray*) and oyster (*black*) culture grounds and observation station for water transparency (*Sta. 22*) near the Hinase Fishermen Union office

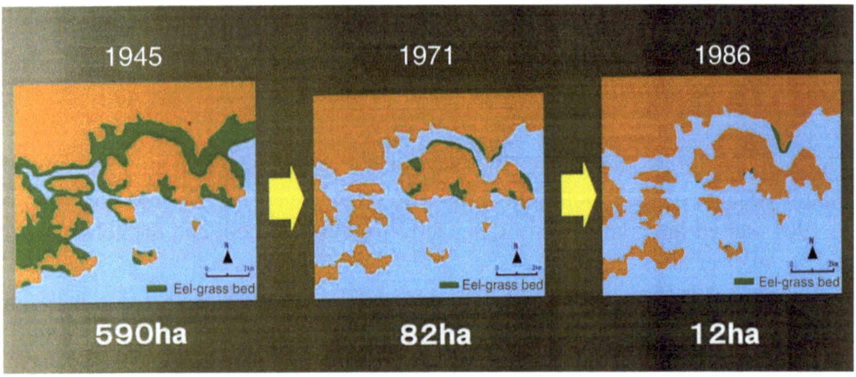

Fig. 3.7 Decrease of eelgrass beds in Hinase fishery ground (from the Okayama Fishery Experimental Station)

Rehabilitation of a eelgrass bed is possible by two means, that is, by sowing seeds or transplanting roots. The fishermen of Hinase Fishermen Union adopted the method of spreading seeds. In May or June, they gather seeds in the eelgrass bed and preserve them in a net under the oyster culture rafts. They select seeds with good quality in October and spread them on suitable areas in November and December (Fig. 3.10).

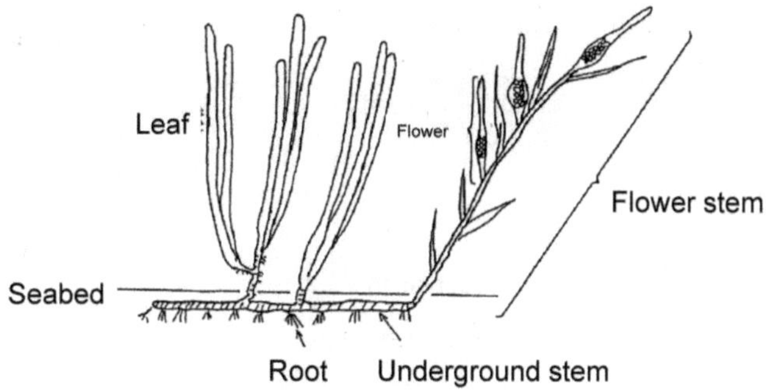

Fig. 3.8 Diagram of eelgrass growth (Naruse 1990)

Fig. 3.9 Life history of eelgrass (Naruse 1990)

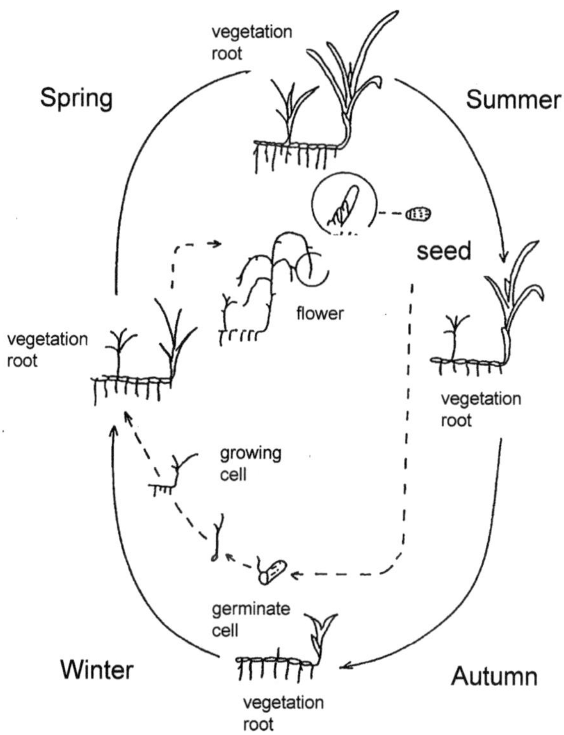

The fishermen spread 150,000 seeds in eight areas (Fig. 3.11) in 1985. After that the number of seeds spread increased, to 2,200,000 in 1988. The area "A" in Fig. 3.11 was an eelgrass bed that had disappeared in 1985 when seeds were spread for the first time. Next spring, a small patch of eelgrass bed was discovered (at area A) that survived until autumn 1986. Area "B," where the seed spreading was carried out in 1986 and 1987, has now become an eelgrass bed. However the eelgrass bed has not formed in area "C" where seeds were spread.

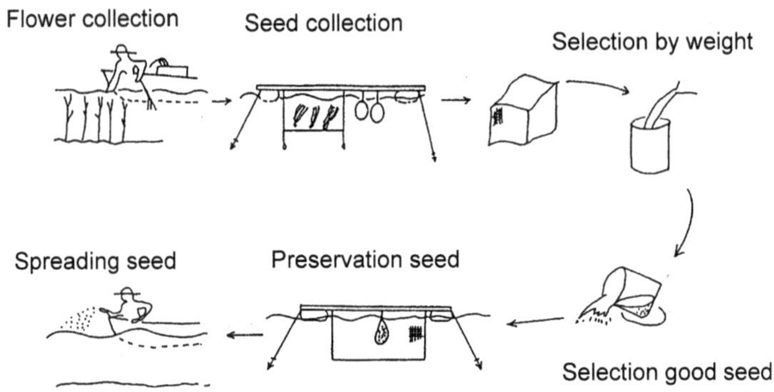

Fig. 3.10 Method for spreading eelgrass seed (Naruse 1990)

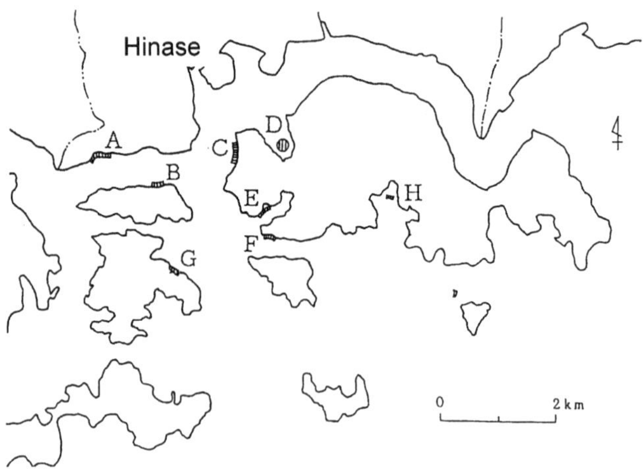

Fig. 3.11 Areas of eelgrass seed spreading (Naruse 1990)

The rehabilitation of eelgrass bed in area "D" succeeded after the improvement of the seabed characteristics by introducing some materials such as oyster shell. At the areas "E–G," rehabilitation succeeded, but it failed at area "H" because sand covered the seabed in this area.

The fishermen have some conclusions on suitable areas for spreading the seeds of eelgrass: (1) areas with a weak tidal current that cannot move the seeds, (2) areas with sandy silt bed where the eelgrass can spread its roots, (3) areas with previous eelgrass beds, and (4) areas with water depth of 0.5–1.0 m at low tide so that sunlight can penetrate to the sea bottom. (5) The establishment of a new eelgrass bed takes a few years (Naruse 1990).

The area of eelgrass beds recovered to 120 ha in 2008, from only 12 ha in 1985, by such continuing activity of the fishermen, and the fish catch of swimming

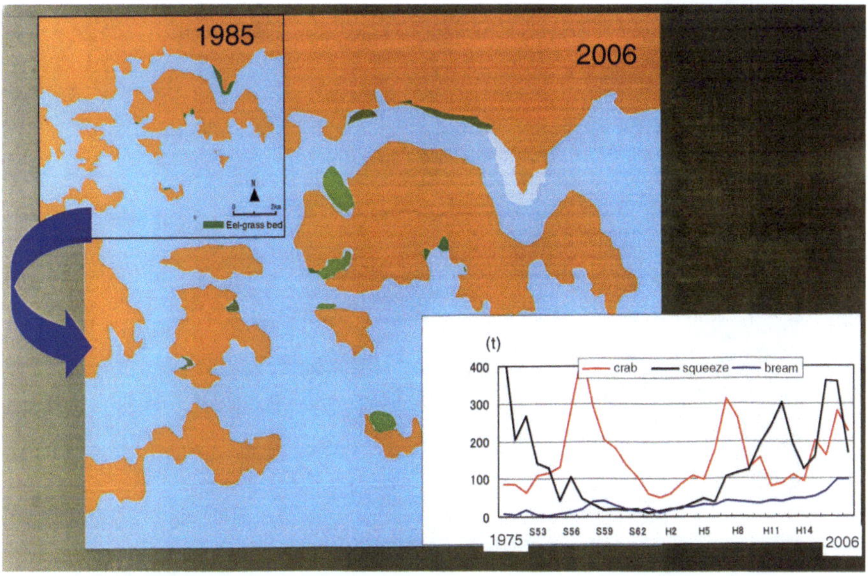

Fig. 3.12 Rehabilitation of eelgrass bed and temporal variation in fish catch by seine net of the Hinase Fishermen Union (from Okayama Prefecture)

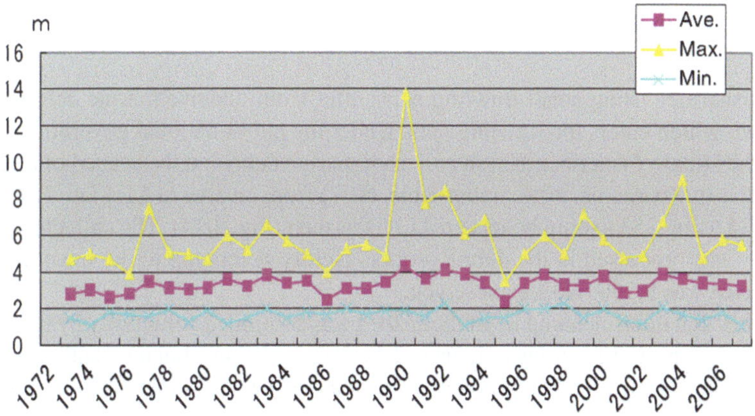

Fig. 3.13 Year-to-year variation in water transparency at Sta. 22 (see Fig. 3.6) (from the Okayama Fishery Experimental Station)

crab, squid (squeeze), and red sea bream by seine net also recovered (Fig. 3.12). Such recovery of the eelgrass bed area may be partially the result of the increase of water transparency in this area (Fig. 3.13).

A manual of eelgrass bed rehabilitation was produced based on their experience.

Fig. 3.14 Marine debris collected from the seabed by trawling net of the Hinase Fishermen Union

3.2.2 *Collection of Seabed Debris*

The fishermen using small trawling nets suffer from massive marine debris on the seabed, which enters the trawling nets. Under the aid of national government funding, the Hinase Fishermen Union collected marine debris on the seabed of a volume of 182.1 m^3 (burnable, 40%; unburnable, 60%) from an area of 11.4 km^2 from 1982 to 1984 using 253 fishing boats and 413 fishermen (Fig. 3.14). The collected marine debris was processed in the debris process factory belonging to the town hall. The total cost was 9,790,000 Japanese yen.

After that time, cleaning up of the seabed was continued voluntarily by fishermen. At the beginning, 12 ton/day of marine debris was collected, but it recently decreased to 5 kg/day. However, the amount of collected marine debris increases after a heavy rain or a passing typhoon.

3.2.3 *Direct Selling*

Half of the fish catch by Hinase Fishermen Union is sold by brokers and another half by direct selling at the Gomi-no-ichi shop, which is operated by the Fishermen Union itself (Figs. 3.15 and 3.16). Hinase is very near to the big cities such as Osaka, and so many customers come to this shop every day.

Fig. 3.15 Direct sales shop of the Hinase Fishermen Union

Fig. 3.16 Inside of direct sales shop

3.2.4 Oyster Culture

Oyster culture is a seasonal fishery during winter, and cultivation of the seabed of the oyster culture ground is conducted by the small trawling net fishermen after the oyster harvest season for conservation of the fishing ground of oyster culture. The accumulated organic matter on the seabed of oyster culture ground is easily decomposed by this cultivation. About 100 young people from China work in the factory for processing the harvested oysters, which is operated by the Hinase Fishermen Union.

The leader of the Hinase Fishermen Union, Mr. Honda (born in 1936), says "We have to develop the sixth industry, which means the combination of the first industry of fisheries, the secondary industry of oyster processing and third industry of direct selling the harvest, for the future of Hinase."

3.3 Catching Herbivorous Fish for Recovering the Dead Marine Forest: Sagara Fishermen Union in Shizuoka Prefecture

There were widespread marine forests of Sagarame and Kajime within an area of 8,000 ha in the Hainan coastal sea in the western part of Suruga Bay, Shizuoka Prefecture, but such extensive marine forests began to disappear about 1985. The phenomena proceeded in 1992, and all the marine forests area had disappeared in 2004 (Fig. 3.17).

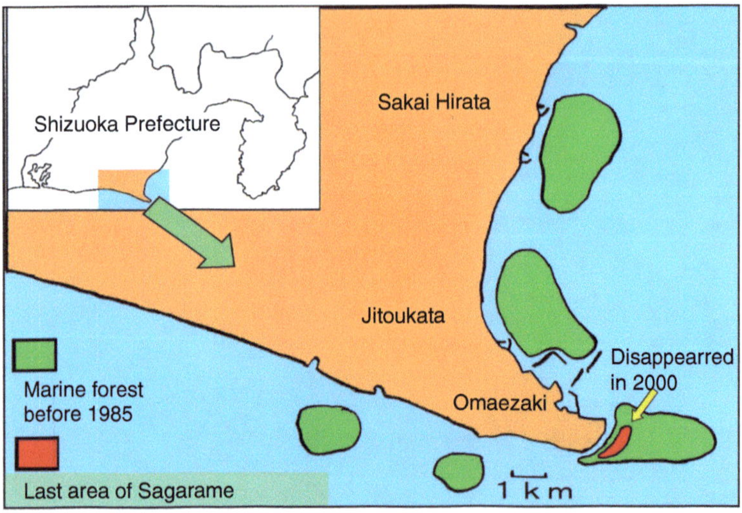

Fig. 3.17 Areas of dead marine forest in Shonan district (Umino 2008)

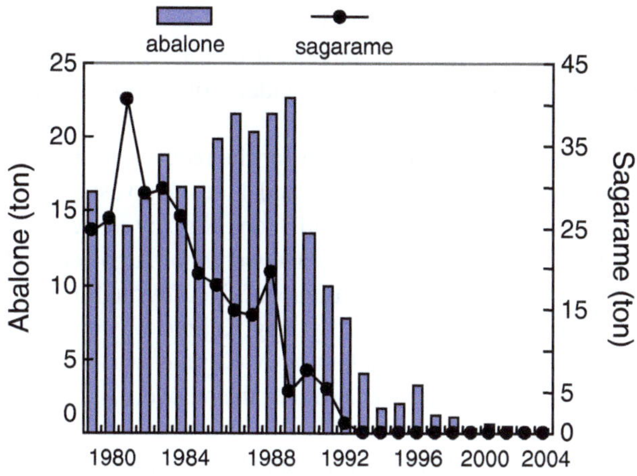

Fig. 3.18 Year-to-year variation in abalone and Sagarame catch in Shonan district (Umino 2008)

The reasons that the marine forest disappeared in this area were (1) decrease of water transparency caused by turbid water from the Ooi River, (2) inflow of weed chemicals (herbicides), and (3) effect of rise in water temperature; however, a final conclusion was not reached. The scientists of the Shizuoka Prefecture Fishery Experimental Station pointed out that grazing by local herbivorous fish (called Aigo by the local people) was largely responsible for the disappearance of these marine forests.

At the same time as the marine forest vanished, abalone, which graze the Sagarame and Kajime, decreased, and the abalone catch, which was 20 ton/year before the disappearance of the marine forests, became zero in 2004 (Fig. 3.18).

Therefore, the Sagara, Jitoukata, Omaezaki, and Yoshida-cho Fishermen Unions established the Hainan association for the recovery of the marine forest in 1996 and began activities for its recovery.

3.3.1 Activities of Recovery by the Marine Forest Association

The Shizuoka Prefecture Fisheries Experimental Station carried out a field experiment of transplanting Kajime from Izu-Shimoda to Hainan in 2009–2010 using concrete blocks. The Hainan association for the recovery of marine forest spread nets around the transplanted Kajime to prevent grazing of Kajime by Aigo. At the same time, the fishermen tried to catch herbivorous fish (Aigo).

Aigo was not sold because it gave off a bad smell and had a poisonous fin; thus, fishermen who caught Aigo released them to the sea. The Hainan association for the recovery of the marine forest began to buy Aigo that were caught by local set nets to prevent them from grazing Kajime.

The money for buying Aigo was defrayed by Shizuoka Prefecture at one million yen per year the first year and was increased to four million yen per year later. Moreover, the Fishery Agency, Japan provided some money for this project in 2001–2005.

The fishermen of the Hainan association for the recovery of marine forest caught 160 kg Aigo using a temporary standing net and 4 tons Aigo using the set net in 2000, 1.5 tons Aigo by temporary standing net and 3 tons Aigo by set net in 2001, and about 30,000 Aigo in both years because the average weight of Aigo was 300 g.

By such efforts, the transplanted Kajime survived through the year, although a part was grazed, and the recruitment of young Kajime was confirmed. Therefore, Shizuoka Prefecture continued transplanting of Kajime with the aid of the Fishery Agency, Japan, and the Hainan Association for the recovery of marine forest continued the catch of Aigo.

At first, covering the whole area of transplanted Kajime by net was tried, but it was abandoned because so many marine biota became attached to the net and the low light environment damaged the growth of Kajime.

The caught Aigo were transformed to fertilizer by the fertilizer company in Shizuoka Prefecture; small Aigo are not caught, but only large Aigo are recently caught.

By such continuous activities, the marine forest area has recovered to 64 ha in 2009, and the abalone catch has also recovered. Trials of releasing abalone juveniles have begun.

3.3.2 *Sagarame: A Kind of Seaweed*

The leader of the Sagara Fishermen Union, Mr. Hagiwara, said that "We want to recover the catch of Sagarame, a kind of Kajime, which was the special product in Sagara Region. People in this region ate Sagarame in the Miso soup or as boiled dish." Sagarame production was the main fisheries of the Sagara Fishermen Union, and about 80% of the union's members engaged in Sagarame production during winter.

Members of the Fishermen Union transplanted cultured Sagarame to offshore areas in early spring, but growth stopped in summer and disappeared in September because of grazing by Aigo. A recovery strategy based on scientific knowledge is necessary.

Mr. Morishita, who was the section leader of the Sagara Fishermen Union, said that "We do not want to catch Aigo to prevent grazing because the marine forest will be maintained when it will have a large area even under the grazing pressure of Aigo." He believes that the present stand is near such a stage. Moreover, he wants to develop a good way of cooking Aigo that people will enjoy.

He said that the greatest result of the association activities was that the union's members began to think about how to sustain their fishing ground for their children and grandchildren through the activities of Kajime transplanting, Sagarame culture, and Aigo catch.

3.4 Oyster Culture and Preservation of Swimming Crab Resource: Tsunemi Branch of the Northern Buzen Sea Fishermen Union in Fukuoka Prefecture

Not only fishery grounds consolidation such as tidal flats and marine forest but also proper fishery resource management is necessary for the creation of Satoumi. The oyster culture and swimming crab resource management in the Buzen Sea, the western part of Suo-Nada in the Seto Inland Sea, Japan, are introduced in this section.

3.4.1 Characteristics of Buzen Sea and Tsunemi Fishermen Union

Broad tidal flats exists in the Buzen Sea, the western part of Suo-Nada in the Seto Inland Sea, Japan, and fishery activities such as small fishing boat trawling for bed fish, gill nets for subsurface fish, cage traps for swimming crab, oyster culture, and so on are popular there.

The Tsunemi branch of the northern Buzen Sea Fishermen Union had 50 full members and 6 submembers in 2008. A submember is able to become a full member 1 year after the start of professional fisheries work. Therefore, the number of young full members, who have come back from the city areas, is increasing now. Most members of the fishermen union engage in oyster culture from autumn to spring and the swimming crab catch in summer to autumn.

3.4.2 Oyster Culture

The first trial of oyster culture in the Buzen Sea was carried out in 1970 when the seaweed culture declined. A few fishermen in Tsunemi branch tried oyster culture using a few rafts and got good results. After that, oyster culture became popular in the Buzen Sea.

Oyster culture was also begun in Yoshida, Sone, and Minoshima branches near Tsunemi branch after 1983, and oyster culture expanded to the whole area in the Buzen Sea (Fig. 3.19).

Accompanying the increase of oyster culture fishermen, “The Association of Oyster Culture in the Buzen Sea” was established in 1999. The association decided to sell cultured oysters with the common name of “A piece of oyster in the Buzen Sea,” and conducted many activities for dissemination of this brand name in larger cities such as Fukuoka in northern Kyushu. Moreover, the association set standards of monitoring methods for the culture ground, healthy processing of cultured products, and selling methods of cultured products to provide safe and cheap cultured oysters to the consumer.

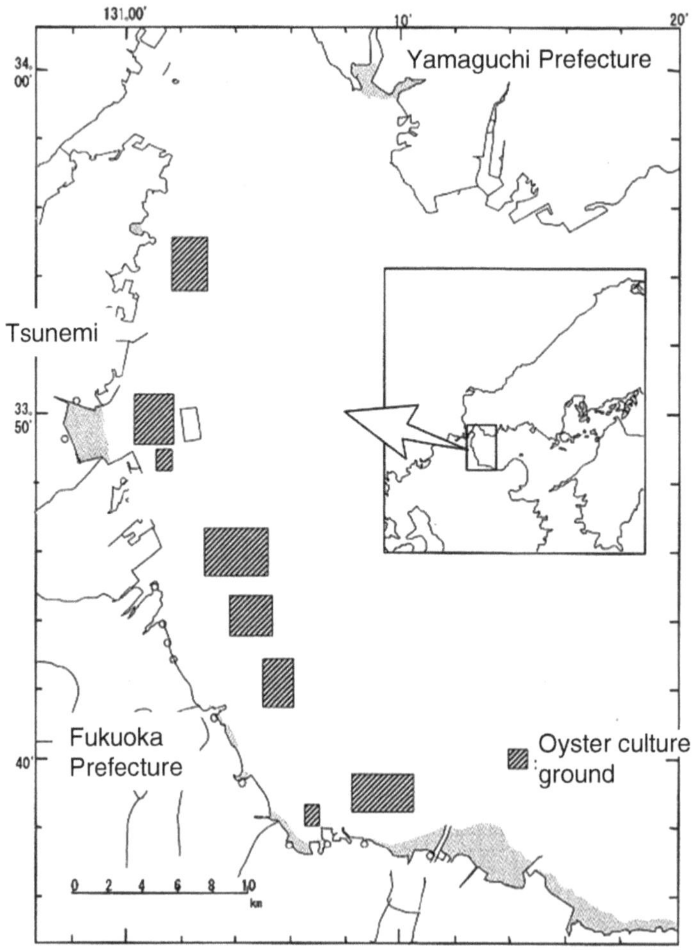

Fig. 3.19 Oyster culture areas in Buzen-sea (Nakagawa 2008)

The association held informative meetings on related themes of oyster culture by inviting an outside lecturer in September every year, when the season of oyster culture begins, and carried out culture ground monitoring to watch for hazardous viruses.

Oyster culture in the Buzen Sea is carried out using bamboo rafts (20 m × 10 m) with about 1,000 hanging ropes to which 13–16 collections of parent oysters, which are bought from Miyagi Prefecture, are attached in March to April every year. The parent oysters are cultured for about a half year in the sea, and the cultured oysters are sold from late November to the next April.

The products are divided according to their sizes, and almost all products are sold with the shell directly to the consumer (Fig. 3.20). The ratio of direct selling and market selling is 6–4.

Fig. 3.20 Direct selling of Buzen oysters

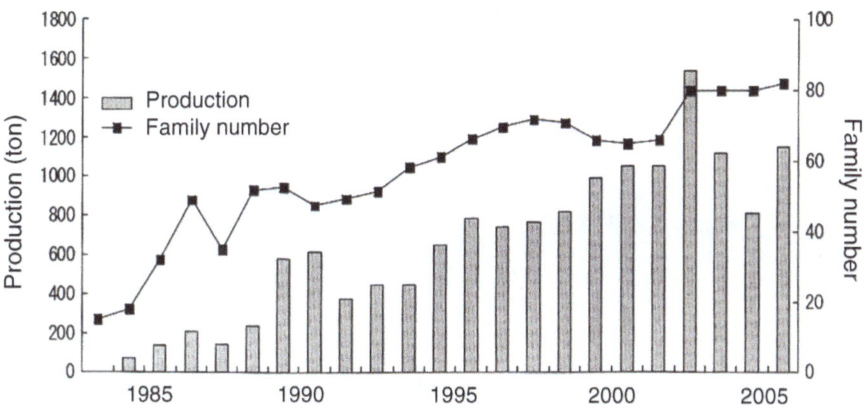

Fig. 3.21 Year-to-year variation in oyster catch and farming family numbers (Nakagawa 2008)

As a result, the products of cultured oysters have been increased since 1995 (Fig. 3.21), and the number of local young fishermen has been increased.

3.4.3 Preservation of Swimming Crab Resource

The young fishermen who engaged in the swimming crab fisheries in summer have been anxious about the decrease in swimming crab catch after 2001 (Fig. 3.22), although they released many young crab juveniles every year. They decided to

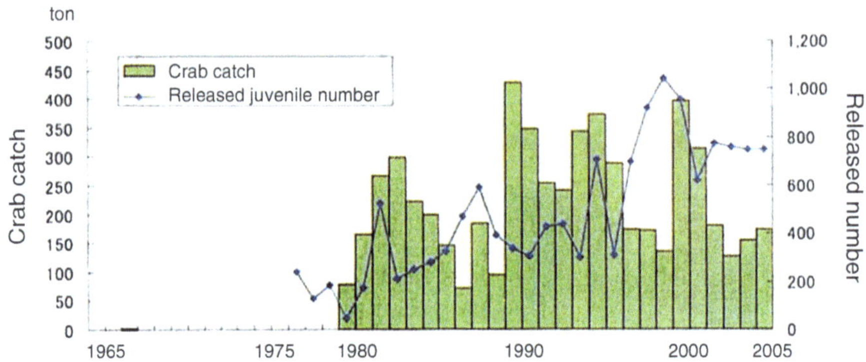

Fig. 3.22 Year-to-year variation in release number of swimming crab juveniles and its catch (Fukuoka Prefecture 2008)

Fig. 3.23 Marked swimming crab purchased with eggs for release (Fukuoka Prefecture 2008)

conduct swimming crab resources management under the suggestion of Fukuoka Prefecture Fisheries Experimental Station.

They began swimming crab resource management, in which they bought swimming crabs caught with eggs in early summer and released them to the sea, from May 2004 onward.

They obtained funding of one million yen in 2004 to buy swimming crabs with eggs from the Fisheries Basic Fund in the Buzen Sea, Fukuoka Prefecture. The purchased swimming crabs were cultured in water tanks on land, marked as "Do not catch" on their back written by magic marker, and were released to the sea on the next day (Fig. 3.23). When the released crabs were caught again in the sea, the fishermen released them at once when seeing the mark on their back.

The price of caught swimming crabs was 500 yen each, which was half the market price of a swimming crab of 16 cm (average size in this season), and it is decided to be the same for all caught crabs irrespective of their size. They asked for cooperation for resources management by local fishermen and buyers in the market; that is, the buyer did not buy a swimming crab with the mark "Do not catch" on their back. The target of 2,000 caught crabs was realized in the middle of July 2004. No marked crab was seen in the market in this year.

The price of bought swimming crab was changed to 400 yen each after 2008 to buy more swimming crabs within the limited budget. Moreover, the budget increased to two million yen after 2005. The catch of swimming crab increased from 127 tons in 2003 to 155 tons in 2004, to 175 tons in 2005, and 206 tons in 2006 as the result of this project (see Fig. 3.22).

This project had a high evaluation from the local people, expanded to the whole area of Buzen Sea, and bought 8,500 swimming crabs with eggs in 2008. However, the local consumer did not know this swimming crab well, and the Fishermen Union tried to give this swimming crab a brand name, "Buzen True Crab," after 1997.

Continued fisheries resources management by local fishermen themselves is very important for the creation of Satoumi.

3.5 Environmental Education: Association of Playing at the Beach in Oita Prefecture

The main players in Satoumi, which is defined as "the coastal sea with high biodiversity and productivity under the human interaction," are fishermen who conduct economic production activities there.

However, the population of fishermen in Japan was only 0.2 million in the year 2000, which is only 0.2% of the total population of Japan (one hundred and ten million at that time). For the creation of Satoumi, the other 99.8% of the people, the citizens, must be included in the Satoumi activities.

Environmental education to teach the importance of the coastal sea is the first step for including the 99.8% of the people in the activities of Satoumi creation. The NPO (Not-for-Profit Organization) "Association of playing at the beach," which has its office at Nakatsu City, Oita Prefecture, the western part of Japan, is well known by its activities of conserving the tidal flats and its advanced environmental education program.

3.5.1 Establishment of the Association

"Association of playing at the beach" was established in July 1999, when the new port development plan at Nakatsu tidal flats (Fig. 3.24) was in progress. The automobile production companies gathered on land in front of the Nakatsu tidal flats, which extend to a length of 10 km along the coast and to 3 km offshore, and

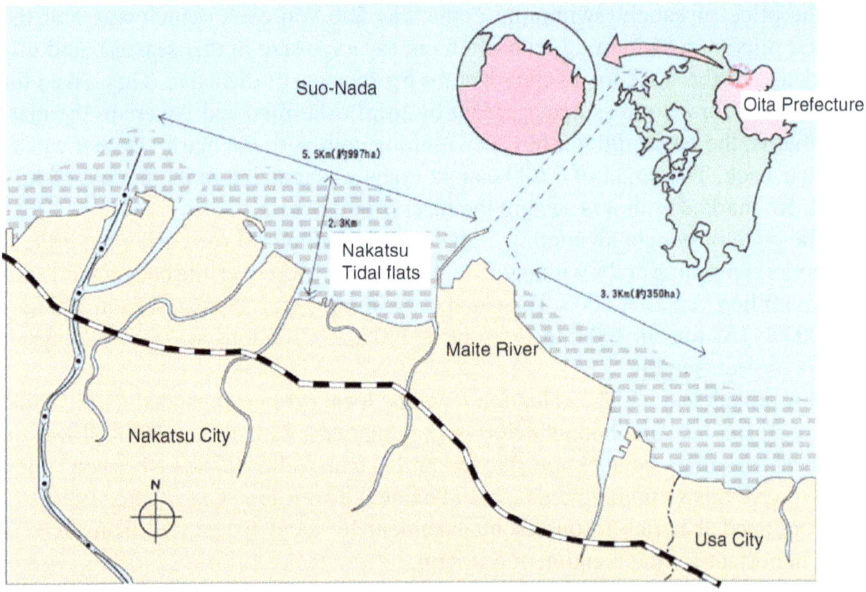

Fig. 3.24 Nakatsu tidal flat (HP of Association of Playing at Beach)

construction of a new navigation route by dredging was planned at the central part of the tidal flats. The local people were anxious about the birds that live and feed on the tidal flats and established the association.

The main player of the association was Ms. Yukiko Ashikaga, who was born in Nagano Prefecture in the central part of Japan, studied animal taxonomy at Ochanomizu University, and moved to Nakatsu City after the employment of her husband (a junior high school teacher) there in 1991 (Ashikaga 2006).

The association first held an observation meeting at the tidal flats with children in July 1999 and discovered juveniles of Japanese horseshoe crab there. The existence of the Japanese horseshoe crab is recorded in the environmental assessment of the port development but was not thought to be important before this discovery. Such observation meetings have been continued every 2 months. The attendance fee is 200 yen per adult (100 yen for students younger than high school). About 200 people attended this observation meeting during the summer vacation.

Nakatsu port was designated as one of the national important ports in November 1999, and it became the duty of Nakatsu port to hear the opinion of local people. Therefore, Oita Prefecture established a committee for environmental management around Nakatsu port in April 2000. Not only the members of "Association of playing at the beach" but also other local people and fishermen joined this committee, and some plans related to the environmental conservation near Nakatsu port were proposed (details are shown later).

3.5.2 *Activities of the Association*

The association carried out beach cleanup and beachcombing four times a year to express their thanks to the beach. The beachcombing is conducted following the manual of JEAN (Japan Environmental Action Network), and 100–300 participants from local junior high schools, senior high schools, universities, and companies attend this activity. The gathered garbage was treated at some cost by the city government at first but is now treated free. The amount of garbage from offshore does not change, but that derived from the land has begun to decrease, which suggests that the continued beach cleaning activity decreases the amount of garbage at the beach.

At the same time, the association carried out research, for example, ecological monitoring around the setback wall at the Maite River estuary (explained later).

The association consults with professionals first, decides how to conduct the work, carries out research itself, and finally consults the professional again. The samples obtained by their research are all available to the public through the virtual museum (explained later). Moreover, they conduct research on the local history of the beach and the coastal sea.

The participants in the research say that “When we attend the research of the association, we can get the fun of much information from the teacher outside, discovering many things, and telling them to the others.” Citizen science is increasing and maturing here.

Ms. Ashikaga conducts several dozen lectures or classes at the nearest elementary schools or high schools in a year.

The association changed to be an NPO in 2006 to obtain project funds from the central and local governments.

3.5.3 *Virtual Museum*

The association has no place for preserving the samples that are collected in their investigations, and thus established the virtual museum in the homepage of the association named “MUSEUM for the association” in 2003. The museum rents the collected samples to the people without a fee. This museum has about 10,000 accesses in a year. The management of this museum is conducted by the husband of Ms. Ashikaga.

3.5.4 *Finances of the Association*

The annual budget of the association is about six million yen, consisting of two million yen from the central and local governments, three million yen from the unofficial funding agencies, and one million yen from donations including the membership fee (1,000 yen per person). However, there is no salary for the staff, including the president.

Ms. Ashikaga says that the activities of an environmental NPO cannot be sustainable if its staff cannot gain a salary from the activities of the NPO. I agree with her opinion.

3.5.5 Preservation of the Maite River Estuary

One of the great contributions of the association is the setback seawall construction at the Maite River Estuary. This construction was conducted as one of the anti-disaster projects for the prevention of storm surge at the Maite River Estuary, where coastal erosion had occurred. To protect the sandy beach for a length of about 200 m, Oita Prefecture bought the land around the beach and constructed the seawall onshore from the beach in 2004, 2 years after the beginning of the discussion. The beach itself is preserved, and the seawall is constructed on land far back from the beach (Fig. 3.25; Seino 2008).

Environmental monitoring has been carried out by the association under funding from Oita Prefecture, showing that the coastal ecosystem, including the breeding of the Japanese horseshoe crab, is not changed.

The slogan of the association is "Satoumi, Satohama, Toyoashihara, and Nakatsu," and the members think that it is most important for the people to continue to relate to the tidal flats for the conservation of the flats.

To tell other people about the richness and enjoyment of the Nakatsu tidal flats, the association conducts activities such as octopus fishing and seaweed culture for children with the help of older fishermen. Moreover, the association cuts bamboo in the mountain area near the Nakatsu tidal flats and reproduces a set net using the bamboo at the flats.

To decrease the distance between the people and the coastal sea, the association began to encourage consumption of local products by the local people.

3.6 Including Citizens in Fishing Activities: Oshima Fishermen Union in Fukui Prefecture

The Komegawaki Branch of Oshima Fishermen Union in Mikuni Town, Fukui Prefecture, has its own fisheries ground along the coast between the Kuzuryu River mouth and the famous sightseeing spot of Tojinbo (Fig. 3.26), where they have caught abalone, turbot, sea algae, and seaweed by continuously taking care of this fisheries ground. However, their fisheries ground has been devastated and the fish catch decreased by the decrease in the number of fishermen and the lack of ongoing care.

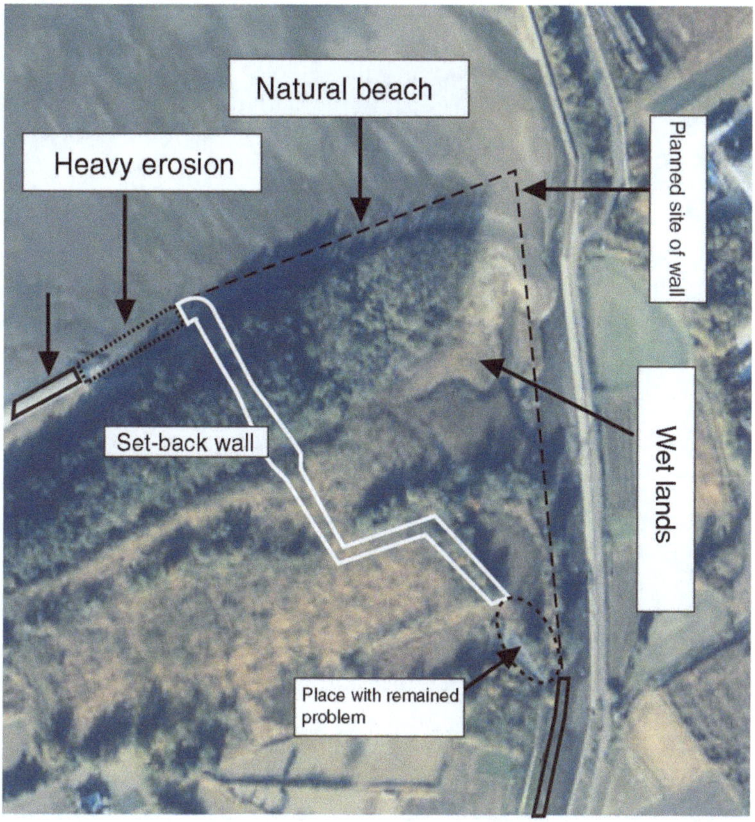

Fig. 3.25 Setback of seawall at Maite River estuary (Seino 2008)

I introduce some activities of Komegasaki Branch of Oshima Fishermen Union in response to such trends in this section.

3.6.1 Activities and Fish Catch of Oshima Fishermen Union

Work is conducted throughout the year by Komegawaki Branch of Oshima Fishermen Union to conserve the bioproductivity and biodiversity in their fishing ground. The marine algae and seaweed fisheries vary seasonally from Haba (a kind of seaweed) and Sugamo (a kind of sea algae) in March, Wakame (a kind of seaweed) in May, Tengusa (a kind of sea algae) in July, Mozuku (a kind of sea algae) in August, and Iwanori (a kind of sea algae) in July. For the harvest of Wakame in May, cutting of the eelgrass is necessary because the eelgrass is an obstacle for

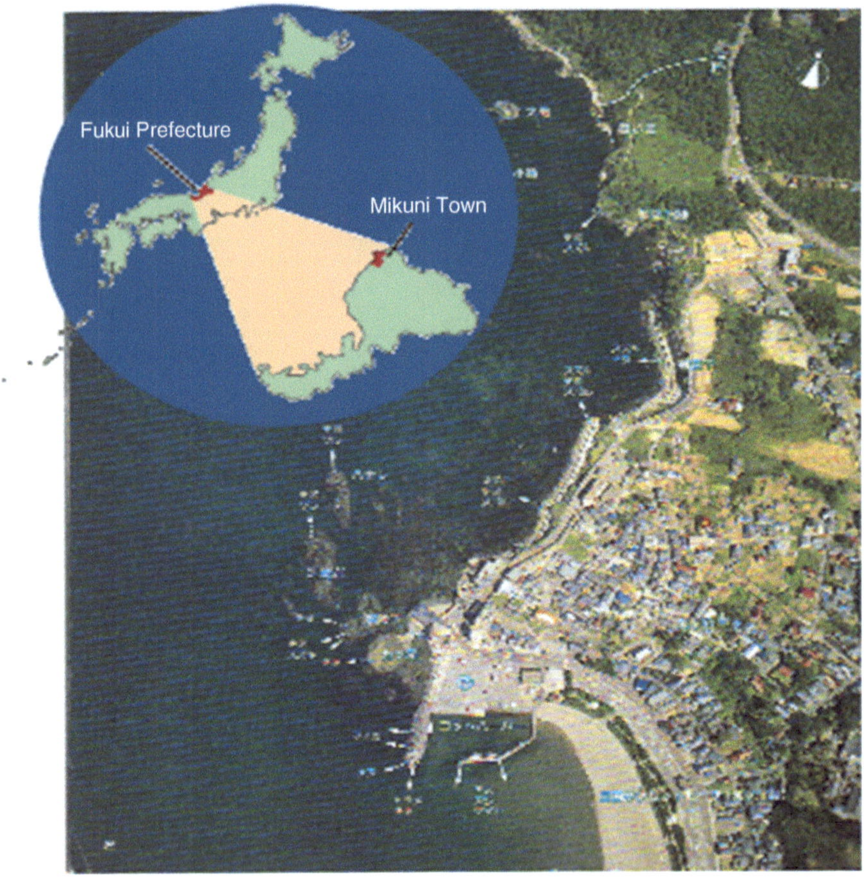

Fig. 3.26 Nearshore fishing ground of Oshima Fishermen Union

Wakame fisheries. To grow Tengusa under full sunshine after Wakame, fishermen must renew the rock surfaces by overturning the rocks. To grow Iwanori in winter, the fishermen must polish the rock surface.

Moreover, periodic beach cleaning is necessary because the marine debris that comes ashore increases, and sand dredging is also necessary because drifting sand buries the fishing ground. The release of urchin juveniles is carried out by hand.

The main workers for these activities were the fisherwomen, but they became old, and there were no successors; that is, 100 fisherwomen in 1985 decreased to 17 in 2009. Therefore, sufficient work for the conservation of the fishing ground became impossible, the ecosystem in the fishing ground changed, the fishing ground is devastated, and the fish catch has decreased.

Fig. 3.27 Komegawaki coast, which Oshima Fishermen Union plans to open to the citizens

3.6.2 Collaboration of Fishermen and Citizens

The deterioration of a fishing ground because of the decrease in the number of fishermen cannot be solved by the fishermen themselves, and they have to think about a new management strategy for their fishing ground that includes the citizens.

As one of such new management options, the Komegawaki Branch of Oshima Fishermen Union decided to open a part of their fishing ground to the citizens; that is, citizens can take sea algae, seaweed, abalone, turbot, sea urchins, and so on there by paying some fee to the Fishermen Union (Fig. 3.27).

3.6.3 Including Citizens

The coastal sea, except the swimming beaches, is thought to be closed to the citizens, and there have been many conflicts between citizens and fishermen related to the catchment of beach biota. How can we open the coastal sea as a field of environmental education and nature experience?

Mr. Y. Mastuda (member of Oshima Fishermen Union) and his wife Ms. N. Matsuda (fisherwoman) established the "Nancy Nature School" to teach children correct knowledge about the coastal sea. Moreover, the husband and wife carry out environmental education, called "Friends of Fishermen," throughout the year on the basis of fishing experiences.

The number of fishermen is expected to continue to decrease in the near future under the present social situation, and we have to establish a new good relationship between fishermen and citizens for Satoumi creation.

3.7 Partial Sale of Fishing Rights: Amino-cho Fishermen Union in Kyoto Prefecture

Many city people say that the mountains are more familiar to them than the sea because the mountains are easier to access than the sea and also the mountains are safer than the sea. One such difference is thought to be caused by the fishing right, by which the fishermen exclusively carry out all fishing activities and city people are not allowed any fishing activity at the sea.

Moreover, the city people have a bad impression about fishermen who obtain large amounts of money by selling fishing rights to factories on reclaimed land. On the other hand, some fishermen unions try to sell part of their fishing right to the city people because the number of fishermen has decreased from year to year.

The trial of partial sale of the fishing right at the Amino-cho Fishermen Union, Kyoto Prefecture, is introduced in this section.

3.7.1 Fishing Right

The fishing right is the right of exclusively carrying out fishing activity in some designated coastal sea area, and it is granted by the prefecture governor. The fishing right is a kind of material right in the legal system, but it is not permitted to transfer the fishing right to another person. The fishing rights are classified into three types: that is, the set net right, the fishing right at a limited area (such as fish culture), and the common fishing right (Iri-ai in Japanese), depending on the fishing methods. Of course, the city people cannot obtain any kind of fishing right.

The present fishing right system was began in 1874 by the Meiji Central Government, which canceled the old fishing custom, asked the fishermen to apply for permission of exclusive use of sea surface area, and tried to take the use fee from them. However, because so many troubles happened throughout the whole country, the Government decided to stop taking the use fee in 1875 and declared that the management of fisheries would be carried out on the basis of old fishing customs and the fishermen must supply the general tax. Moreover, the government established the rules of fishermen unions and the principal law of fisheries in 1900, and began to manage the fishing problems.

However, the people demanded the innovation of this fishing law because of the bad arrangement of fishing right definition and the fisheries union system, and therefore the government innovated the fishing law and established the so-called Meiji Fishing Law in 1909, establishing the new fishing right, the fishing permission system, and the fishing control system by this new legal system.

The modern revision of the fishing legal system took place in 1949 after the Second World War at the same time as the revision of the agriculture legal system. The Meiji Fishing Right vanished, and the new fishing right was given to the local fishermen who lived near the coastal sea. The new fishing legal system defines the fishing right as "fishing activities by the local fishermen for their life" and the fishermen are defined as "the people who conduct fishing activities for their life."

3.7.2 One-Day Fisherman Card in Amino-cho Fishermen Union

Amino-cho Fishermen Union (198 members) is an assembly of small unions with several fishermen, which exists at Amino Town in Kyoto Prefecture. Shimazu Branch is one of such small unions, with 12 members, and it sells a "One-day fishermen card" (Figs. 3.28 and 3.29) during a limited time in summer at the price of 2,500 yen per day (for diving) or 500 yen per day (for hand-line fishing) at a limited area extending 700 m along the beach and 100 m offshore (Figs. 3.28, 3.30, and 3.31) at the Kotobiki-Hama coast, which is very famous as a singing beach.

With this card, city people can catch abalone, turbo, bivalves, sea algae, seaweed, and so on at this beach. The fishermen with the fishing right sell a part of their resource to the city people without the fishing right, similar to the right of fishing sweet (freshwater) fish in Japanese rivers.

This trial was conducted every summer for 1.5 months from 1 July to the middle of August, and the Shimazu Branch obtained an income of about 10,000 yen in 2007, about 100,000 yen in 2008, with 39 diving and 19 hand-fishing cards, and about 190,000 yen in 2009 with 72 diving and 17 hand-fishing cards. City people using the card must respect the local rules on permitted catch size of more than 2 and 10 cm for turbo and abalone, respectively, and be responsible for their own safety.

The customers for this one-day fishermen card are increasing, and many repeaters come back to Kotobikihama beach to buy the one-day fisherman card. One father who attended this project was very glad to show the octopus caught by his diving to his children.

Environmental monitoring by the Kyoto Fisheries High School suggests that no distinct environmental change has been caused by this project in the fishing ground where the one-day fishermen card is sold.

The beginning of this project was not easy because it was necessary to get the agreement of the 12 members of Shimazu Branch who have the fishing right at this fishing ground and the agreement of the officers of Amino-cho Fishermen Union. Mr. S. Matsuo (born in 1963), who is the leader of this project, says that it took

Fig. 3.28 Advertisement of one-day fisherman card in Amino Fishermen Union

Fig. 3.29 Certification of one-day fishermen in Amino Fishermen Union

Fig. 3.30 Kotobiki beach with Taiko-hama at the central part

Fig. 3.31 Taiko beach with Ooiwa in falling snow

3 years to obtain all the agreements. The main reason of the objection of local fishermen to this project was the anxiety of losing their fishing right by this project.

3.7.3 Conservation Association of Singing Sand at Kotobikihama Beach

Mr. S. Matsuo, who is a member of Shimazu Branch, and is the leader of Asobiura village beach project by the fishermen of small boat trawling and has a small hotel, continues the conservation activity of Kotohikihama Beach, which is very famous as the beach with the singing sand. Such activity is the basis of the one-day fishermen project.

Kotobikihama is very famous with its singing sand, having a coastline length of 1.8 km, and 0.2 million sightseeing guests visit each year (Figs. 3.28 and 3.30). The parking area for the sightseeing guests is maintained by the local Asobiura village society, which sells the one-day fishermen card and solves problems in the fishing ground for one-day fishing.

"The conservation association of singing sand at Kotobikihama Beach" was established in June 1987 by the local people and has conducted beach cleaning, the exhibition of landed materials at the beach, the symposium, and wood planting behind the beach. The first "National Symposium on Singing Sand" was held in 1994 here by the association, and the association succeeded in recovering the beautiful beach in a short time after the Nakhodka oil spill accident in 1997 with the help of 12,700 volunteers from all over Japan.

On the basis of the association activities, the Amino-cho Town designated the Kotobikihama Beach as a special conservation area in 2001 and decided that smoking, fireworks, camping, and cooking at the beach are prohibited because of their bad effects on the singing sand.

Mr. S. Matsuo says that the activities of the association was very useful for persuading the old people to agree to the one-day fishermen system because the old fishermen can understand the possibility of young fishermen such as Mr. S. Matsuo from the activities of the association. Especially, the Nakhodka accident suggested to the old fishermen that the good condition of the coastal sea could not be conserved without the cooperation of city people as volunteers.

The origin of the idea of one-day fishermen exists in the concept of how to build a new relationship between the fishermen who live at the coastal sea and the city people who visit there. Mr. S. Matsuo says that he wants to manage the Kotohikihama Beach under an association by the local people, city people, and local government, to publish the one-day fishermen card not only during summer but also during the whole year, and to establish a co-production system by the local fishermen who catch sea algae, seaweed, and bivalves and visiting city people.

Such trials by the Amino-cho Fishermen Union are the first step of co-production of fishermen and city people to promote the creation of Satoumi, not only by the fishermen but also by the city people.

3.8 Integrated Management of Mountain, River, and Coastal Sea: Fushino River in Yamaguchi Prefecture

Fushino River is a middle-sized river that flows from the northern mountain area at a height of 500–700 m to the southern coastal sea area of Yamaguchi Bay in the central part of Yamaguchi Prefecture; its catchment area is 322 km^2 and its length 30.3 km (Fig. 3.32).

The population in the watershed area of Fushino River increased from 150,000 in 1965 to 170,000 in 2000. The ratio of the sewage treatment area increased from 10% in 1985 to 67% in 2001. There is no water pollution factory in the catchment area of Fushino River.

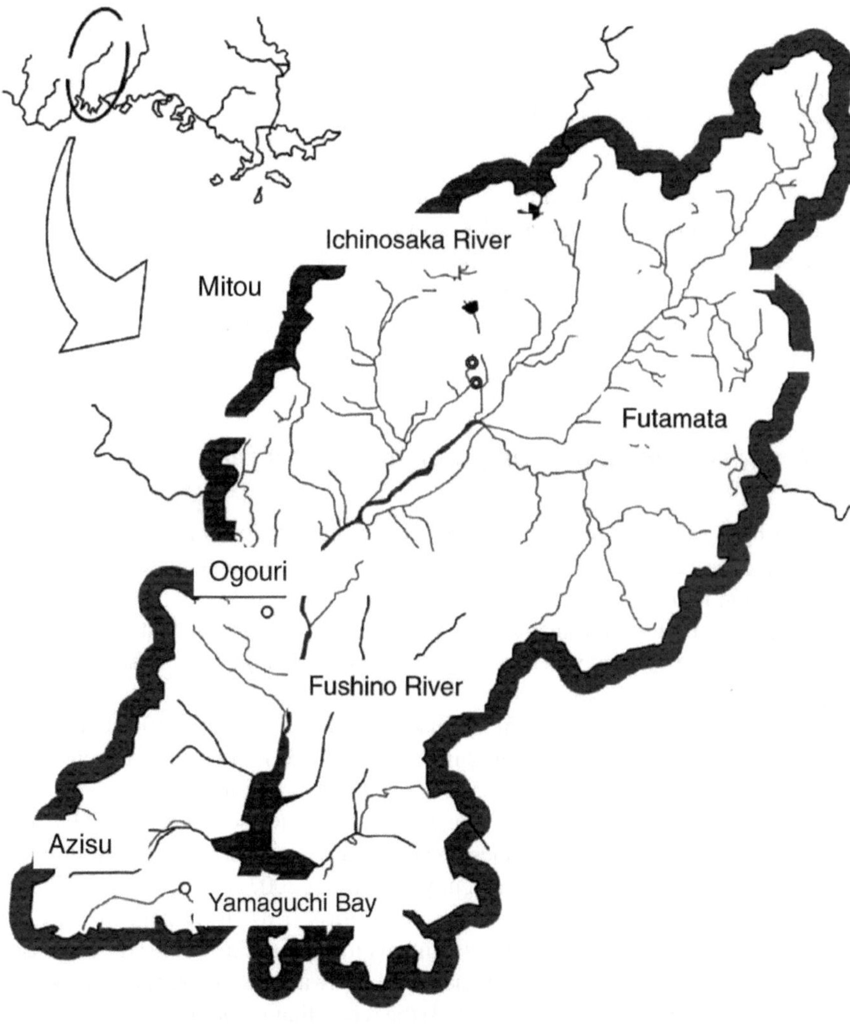

Fig. 3.32 Catchment area of Fushino River (Ukita 2007)

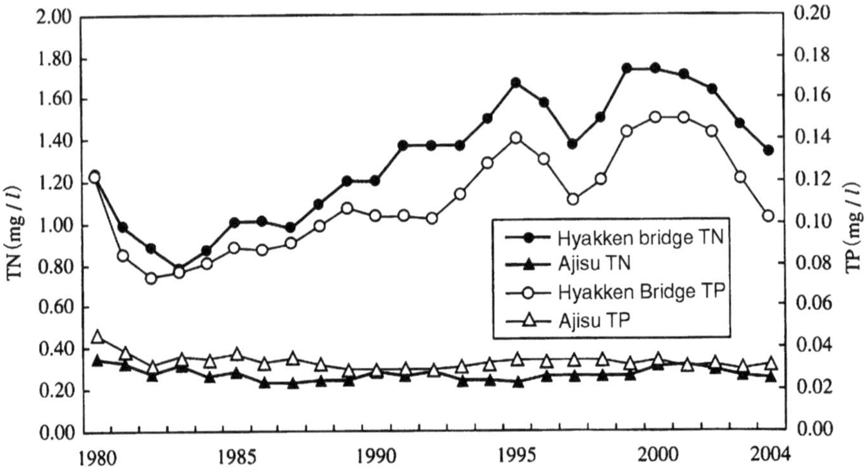

Fig. 3.33 Year-to-year variation in total nitrogen (*TN*) and total phosphorus (*TP*) concentration at the mouth of Fushino River and off Azisu (Ukita 2007)

Fushino River has many weirs for drinking water and water for agriculture, but has only two dams, that is, Ichinosaka Dam with a catchment area of 6.7 km^2 and a water volume of 1.49 million m^3, constructed in 1983, and Aratani Dam with a catchment area of 8.1 km^2 and volume of 5.2 million m^3, constructed in 1989. The ratio of watershed area by these two dams to the whole catchment area of Fushino River is only 4.6%.

3.8.1 Environmental Changes in Yamaguchi Bay

Total nitrogen (TN) and total phosphorus (TP) concentrations in the Fushino River estuary increased from 1982 to 1999 but decreased after that time. TN and TP concentration in Yamaguchi Bay decreased after 1980 (Fig. 3.33). The variation in nutrient concentration in Yamaguchi Bay is mainly controlled by that in Suo-Nada, which decreased in recent years, because the water discharge from Fushino River is very small (Yanagi and Ishii 2008).

The eelgrass beds in Yamaguchi Bay had an area of 720 ha in 1950 but decreased after that time and nearly disappeared in 1990 (Fig. 3.34). The beds recovered after 1990 and increased to 153 ha in 2005 (Fig. 3.34). The reason that the eelgrass beds decreased is thought to be that the diameter of suspended sediments from Fushino River became small and the particles could be easily resuspended, resulting in the decrease of water transparency in recent years (Ukita 2007).

The reasons for the recent recovery of eelgrass beds in Yamaguchi Bay are thought to be that (1) the suspended sediments load from Fushino River decreased because of agricultural land management, (2) the water transparency increased in Suo-Nada (Yanagi and Ishii 2008), and (3) the rehabilitation project of eelgrass beds

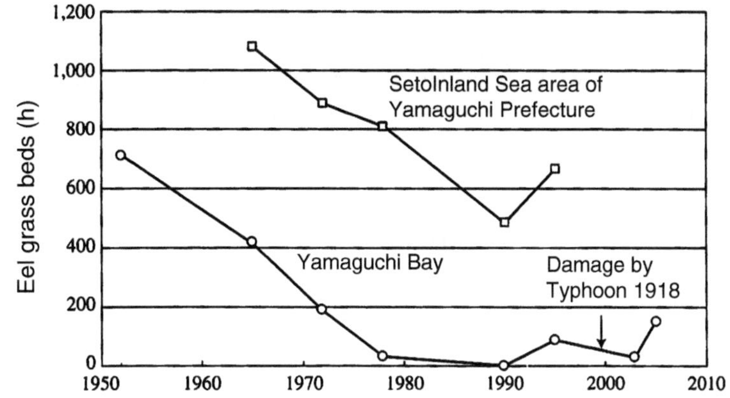

Fig. 3.34 Year-to-year variation in area of eelgrass beds in Yamaguchi Bay (Ukita 2007)

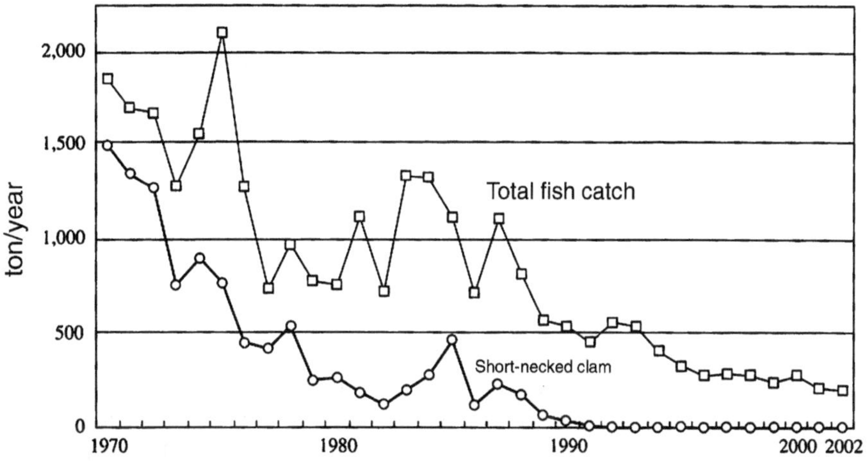

Fig. 3.35 Year-to-year variations in fish catch and short-necked clam catch in Yamaguchi Bay (Ukita 2007)

was initiated by Yamaguchi Prefecture. Therefore, the eelgrass beds in Suo-Nada also increased in recent years (Fig. 3.34).

The fish catch in Yamaguchi Bay rapidly decreased and that of short-necked clam became nearly zero after 1991 (Fig. 3.35).

3.8.2 Collaboration of People in the Catchment Area

Fushino River has been familiar to the local people, and there are many NPOs (Not-for-Profit Organization), such as "The conservation association of root area of Fushino River," "The environmental conservation association of Yamaguchi Prefecture," "The raft group in Fushino River," "The conservation association of fire worm in Ootono," "The healthy growth of young people in Ootono," "The cleaning up association of Sijyu-Hachise River," "The Yamaguchi central woods union," "The horse chestnut woods association," "The rehabilitation association of Fushino River," "The Fushino River Fishermen Union," "The Yamaguchi Prefecture Branch of Japan birds association," "The Yamaguchi Horseshoe Crab Research Group," "Yamaguchi Fishermen Union," and so on. These groups were very active around 2002.

The trigger of collaboration of these NPOs was the activity of "The conservation association of root area of Fushino River," which was started in February 2001. The plan for the industrial sewage treatment plant at Futamata area near the root area of Fushino River (shown in Fig. 3.32) was opened in 2001, and more than 12,800,000 yen were donated to the association from more than 10,000 people in the whole of Japan for objections to this plan. The association bought 4 ha land in the planning area and donated it to Yamaguchi Prefecture, which arranged the natural park there.

The fishermen unions around Fushino River and Yamaguchi Woods Union began their collaboration being inspired by such activities of "The conservation association of root area of Fushino River"; that is, the land people went to the coastal sea for cleaning up the beach and the sea people went to the mountain for planting woods. "The exchange association for the rehabilitation of Fishuno River" was established in 2002, and it has continued its activity up to the present.

Activity to protect the fire-worm at Ootono area in the Ichinosaka River (see Fig. 3.32), which is in the middle of Fushino River, has been conducted for a long time. Ichinosaka River was designated to be a National Natural Monument as the site of the Genji-fire-worm in 1935 and was designated as one of the beautiful landscapes of Japan in 1964. The river bank of Ichinosaka River for fire-worm, which was constructed in 1972 after the big disaster by a typhoon, is very famous as one of the successful river banks harmonizing with its environment. The boys and girls of Ootono elementary school continued the activities of protecting the larvae of fire-worm and cleaning up the river water. As a result, many fire-worms fly in every May and many people visit the Ichinosaka River. The Ministry of Construction, Japan gave a special award to such activities of the people in the Ootono area in 1987.

Yamaguchi Prefecture established the "Special Tax for the Conservation of Forests" in 2005 and collects 500 yen per year per person and 1,000–40,000 yen per year per company, depending on its capital, using such money for the management of forests in Yamaguchi Prefecture.

3.8.3 Basic Plan on the Healthy Catchment Area

Yamaguchi Prefecture set up the "Committee of Promoting the Good Watershed in Yamaguchi" in 2002 based on funds from the Ministry of Environment, Japan. The principles of this committee are as follows:

1. Clean water flows in rivers with plenty of green surrounding them, and many kinds of biota live there.

2. People live there with happiness, proud of their culture that is based on nature.
3. Products from forests, lands, rivers, and seas are exchanged and people also interact there.
4. People in areas of the upper, middle, and lower streams collaborate to attend activities for establishing happy areas and healthy watershed areas based on gratitude for each other.

3.8.4 Rehabilitation Activities of Tidal Flats at the Mouth of Fushino River

Yamaguchi Prefecture decided to carry out "The rehabilitation of tidal flats at the mouth of Fushino River" as one of the important acts based on the "Basic plan on the healthy catchment area." It used the umbrella of the "Rehabilitation of Nature Law," which was enacted by the central government of Japan in January 2003. The legal guides to conduct the rehabilitation projects are based on the decision of the Rehabilitation Committee whose members are the residents, scientists, an NPO, and policy makers. The tidal flats at the mouth of Fushino River are well known as nationally important wetlands because these flats were selected as 1 of 500 important wetlands in Japan.

The rehabilitation committee of tidal flats at the river mouth of Fushino River was started in August 2004 with 8 scientists, 13 volunteers, 18 NPO members, 14 local policy makers, and 4 national policy makers. The principle of this committee is "In order to realize high biodiversity at the mouth of Fushino River, many kinds of people attend to conduct the adaptive management based on the co-production of industry, academy, government, and people." For the moment, the target was defined to recover short-necked clam, which is very familiar to the people, at the tidal flats of Fushino River. Many kinds of actions based on the zoning shown in Fig. 3.36 have been conducted to realize this aim.

The reasons for the decrease of short-necked clams in the Fushino River estuary are said to be (1) decrease of medium-diameter sediments in the bottom, (2) massive deposition of oyster shell, (3) decrease of nutrients in the mud, (4) increase of hardness of mud, and (5) predation by ray-fish, but the main reason has not been clarified.

The committee began cultivation of the bottom mud at the tidal flats in 2005, with the construction of bamboo protectors preventing predation by ray-fish and rehabilitation of the eelgrass bed. The survival rate of short-necked clam was examined from early June to middle September in 2005 using a 50-cm^2 quadrat at five stations shown in Fig. 3.37 (Sekine and Ukita 2008). The survival rate was highest at Stn. 2 (Fig. 3.38a) with coarse mud and a net protector and at Stn. 1 (Fig. 3.38b) where ray-fish could not come. These results suggest that the short-necked clam will recover at the Fushino River estuary if we can prevent predation by ray-fish.

The marine environment of Yamaguchi Bay has recovered by expansion of the eelgrass beds by the cooperation of many associations in the watershed of Fushino

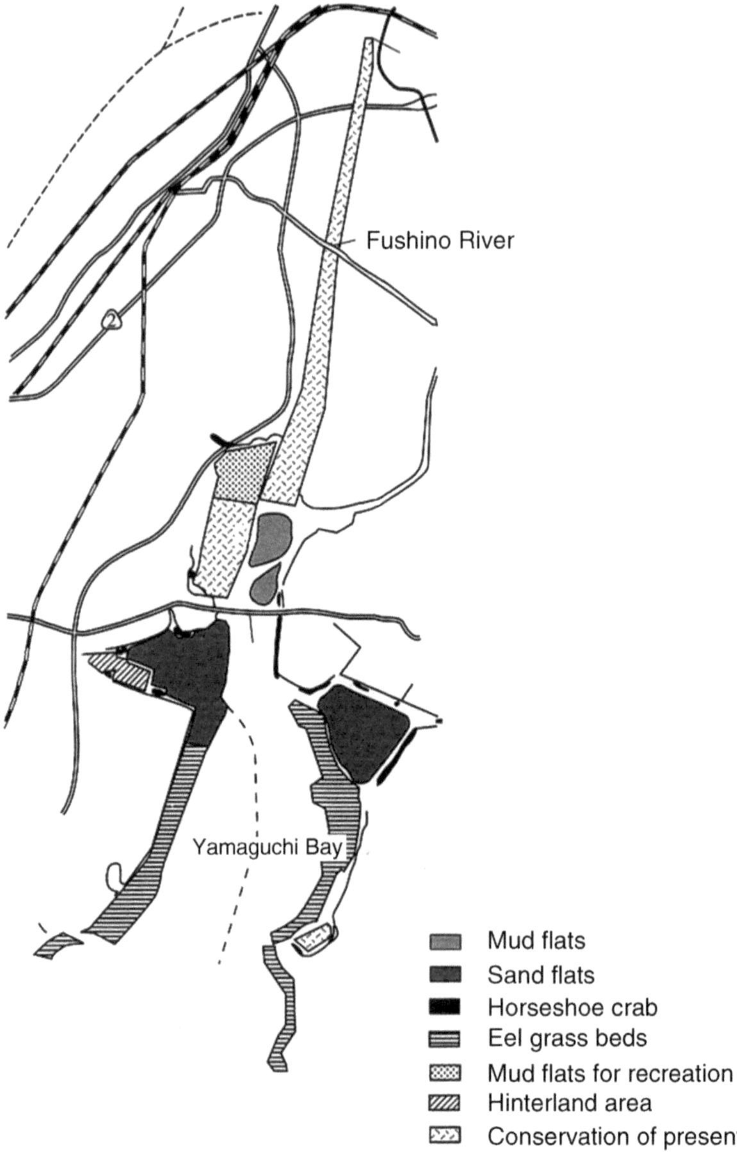

Fig. 3.36 Zoning of Fushino River and Yamaguchi Bay (Ukita 2007)

River, and the short-necked clam is recovering in the Fushino River estuary under suitable human intervention.

The continued activities of such associations are expected in the future, and thus Satoumi will be created in the Fushino River estuary.

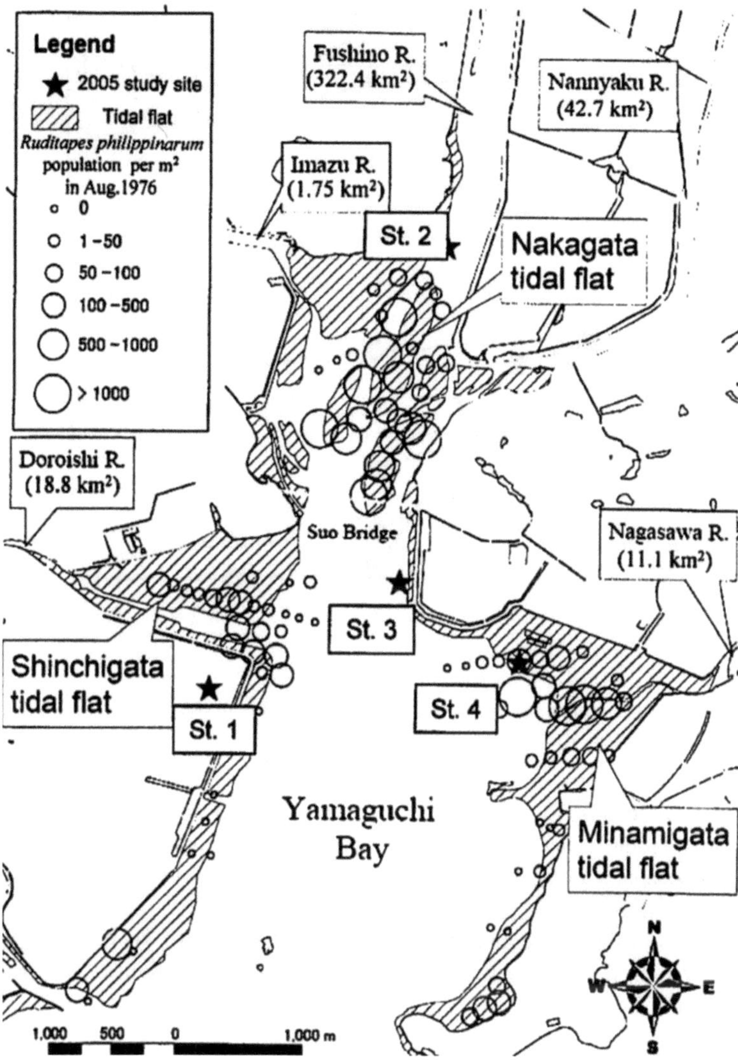

Fig. 3.37 Density distribution of short-necked clam in 1976 (*circles*) and sampling points in 2005 (*stars*) at the tidal flat (*slanted lines*) of Fushino River mouth (Sekine and Ukita 2008)

3.9 Coastal Management: Kashiwajima Island in Kochi Prefecture

Many kinds of activities are undertaken by different stakeholders in the coastal sea, and conflicts sometimes occur among such different stakeholders. The local or central government usually manages such troubles, and the fisheries section plays the

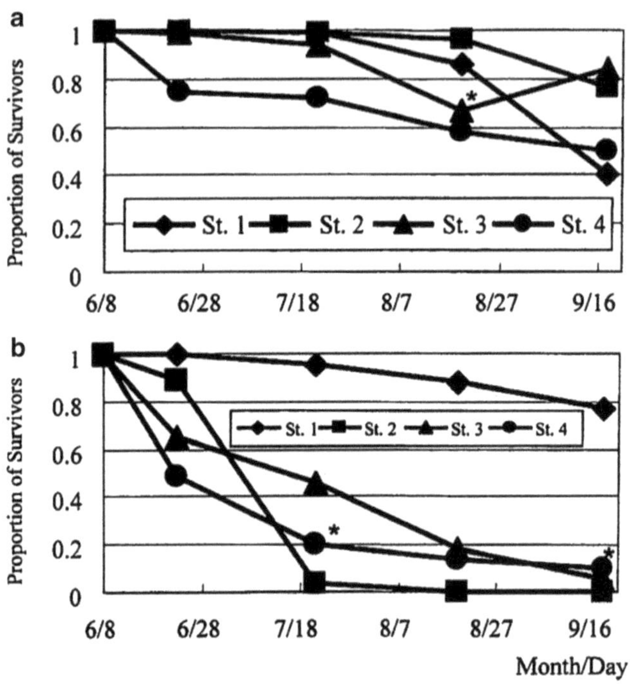

Fig. 3.38 Change of remaining short-necked clam numbers with protective net (**a**) and without protecting net (**b**) for ray-fish (Sekine and Ukita 2008)

important role because the main right in the Japanese coastal sea is the fishing right. However, in recent years, sports fishing, scuba diving, whale watching, and so on have become very popular in the coastal sea, and such activities lead to new troubles with fishermen.

What kinds of management are possible at the time of such troubles? One example of such management at Sukumo Bay in Kochi Prefecture is introduced in this section.

3.9.1 Establishment of Kuroshio Feeling Center

Kochi University negotiated with the Ministry of Education on the establishment of a marine research center at Kashiwajima Island in Sukumo Bay in 1996 but it failed. The people of Kashiwajima Island were disappointed to hear this news. Dr. Masaru Kanda, who wrote his master's thesis in the Agriculture Department of Kochi University based on fieldwork at Kashiwajima Island, received his doctoral degree from the University of Tokyo, and worked as a P.D. (postdoctoral course) scientist in Kochi University at that time, proposed to the people to establish a civil

Fig. 3.39 Kashiwajima Island. *Large circles* (50 m diameter and 20 m depth) are tuna culture nets; the *small circles* are bream culture nets

or prefectural research center at Kashiwajima Island. He moved alone to Kashiwajima Island in April 1998, rented a vacant room in the Kashiawajima Junior High School from Otsuki town, and opened a preparative room for the Kuroshio Research Center.

Dr. Kanda continued negotiations with the local people and held the first Satoumi Symposium at Kashiwajima Island for a 4-km distance around the island and population of 550 (Fig. 3.39) in 2000, with 400 participants, inviting the Governor of Kochi Prefecture, Mr. Hashimoto, and the president of Kochi University, Dr. Tatsukawa. His definition of Satoumi is "coastal sea where people and nature coexist because people gain the ecological service from the nature and people conserve the nature at the same time."

On the basis of this symposium, an NPO (Non-Profit Organization), the Kuroshio Feeling Center, was established in October 2002. The president was Dr. Tatsukawa.

3.9.2 Activities of Kuroshio Feeling Center

The Kuroshio Feeling Center considers Kashiwajima Island itself as the field museum with its transparent seawater and about 1,000 species of fish, and has carried out the following three activities.

The first activity is for the visitors to feel the nature of Kashiwajima Island by investigation and experience. Dr. Kanda wants the visitors not only to experience nature but also to feel nature. He thinks the real feeling of nature is very important for its understanding, and he tries to provide intensive field study to a small number of visitors. The center receives about 500–1,000 visitors every year, and they experience the nature of Kashiwajima Island through field study, summer school, eco-tours, scuba diving, and so on.

The second activity is to help the people's life in Kashiwajima Island by using the nature of the island; that is, the center discusses with the island people how to use the natural resources of their island in a sustainable manner for more than 40,000 sports fishermen and divers per year. Local fishermen consider that "the sea belongs to them" and complain "we cannot catch the squid due to the activities of divers" and "divers are the obstacle for fishing." On the other hand, local divers claim that "coastal sea belongs to all people" and "divers use the coastal sea with great care." The center thinks the mutual understanding of local fishermen and local divers is most important and promotes the project of installing egg beds for squid around the coast of Kashiwajima Island, using unneeded trees from the forest, by the cooperation of the local fishermen and local divers for the coexistence of both sectors. Moreover, the local fishermen and local divers decide the designation of diving areas, prohibiting times and areas for diving after their discussion. The center cooperates to "rehabilitation association of island activity in Kashiwajima Island" for the opening of the local market three or four times per year. Dr. Kanda wants to establish a new marine industry at Kashiwajima Island through such activities.

The third activity is environmental monitoring such as reef check, seaweed bed investigation, fish check, and so on with the aid of local divers and scientists from Kochi University (Kuroshio Feeling Center 2006).

Such contributions of the center have been reported in 10- to 15-min segments as NHK TV local news every month since 2004, and as one-page special news in the evening version of *Kochi Newspaper* every month since 2007.

3.9.3 Budget of Kuroshio Feeling Center

The management of the center is handled by a few staff in addition to Dr. Kanda, who is the director, and the annual budget of about 15,000–30,000 thousand yen is supported by funds from the Toyota Foundation, Japan Foundation, Zerox Foundation, the Ministry of Infrastructure and Transportation, the Ministry of Environment, the Ministry of Education, Japan, and so on. However, such funds are allocated for use by the projects and cannot be used for personal reimbursement. Personal salaries are supported only by the membership fees (personal, 3,000 yen per year; company, 200,000 yen per year). As the present number of members is about 500, only one person can be hired by such fees.

Dr. Kanda wants to obtain money for hiring new staff by executing more eco-tours, but it is not so easy for the center to expand such an enterprise.

3.9.4 Present Problems

The present problems of the center are (1) how to obtain money for hiring new staff, (2) how to improve the relationship between the center and the local government of Otsuki town, (3) how to further improve the relationship between local fishermen and local divers, (4) how to solve the marine pollution problem resulting from tuna culture, which has become very popular in recent years, and so on.

3.9.5 Satoumi Action in Kashiwajima Island

The center wants to establish "Kashiwajima Satoumi Law" now. Some bad effects have appeared in the coral reefs and marine ecosystem around Kashiwajima Island because of the increase of visitors, and friction has developed because of the increase of garbage left by visitors. On the other hand, the economics of Kashiwajima Island depend on such visitors.

The center wants to establish a new relationship between the local people and the visitors by defining a new local rule, that is, "Kashiwajima Satoumi Law," and wants to change Kashiwajima Island from the traditional sightseeing island to a new sustainable environmental island.

The center wants visitors to understand well the life of local people at Kashiwajima Island, especially in reducing water use because the water resource in Kashiwajima Island is very limited, and wants the local people to sustain the rich nature of Kashiwajima Island, which has been transferred from their ancient ones, to pass it on to their grandchildren. Many kinds of fishing such as line fishing, deep-sea coral reef catches, and seine net for tuna have been lost because of the disappearance of fishery resources, which may in turn be the result of overfishing. Diving itself may be disappearing in the future because of overuse of the coastal seas of Kashiwajima Island.

The center claims that all people must be careful to sustain the natural resources in Kashiwajima Island to sustain life there.

A similar trial is underway at Shiraho Town in Ishigaki Island, Okinawa Prefecture (http://www.sa-bu.com/).

3.10 The Coastal Sea Area Around Ikishima Island as a Marine Protected Area (MPA)

It is well known that the indigenous flora is well preserved at Chinjyu-no-Mori (God's Forest) in Japan. The local people have not touched the trees in Chinjyu-no-Mori because they believe that the protecting god of the local village lives in Chinjyu-no-Mori. For example, trees with evergreen broad leaves, such as the camphor tree and pasania, are well preserved at Chinjyu-no-Mori in the western part of Japan.

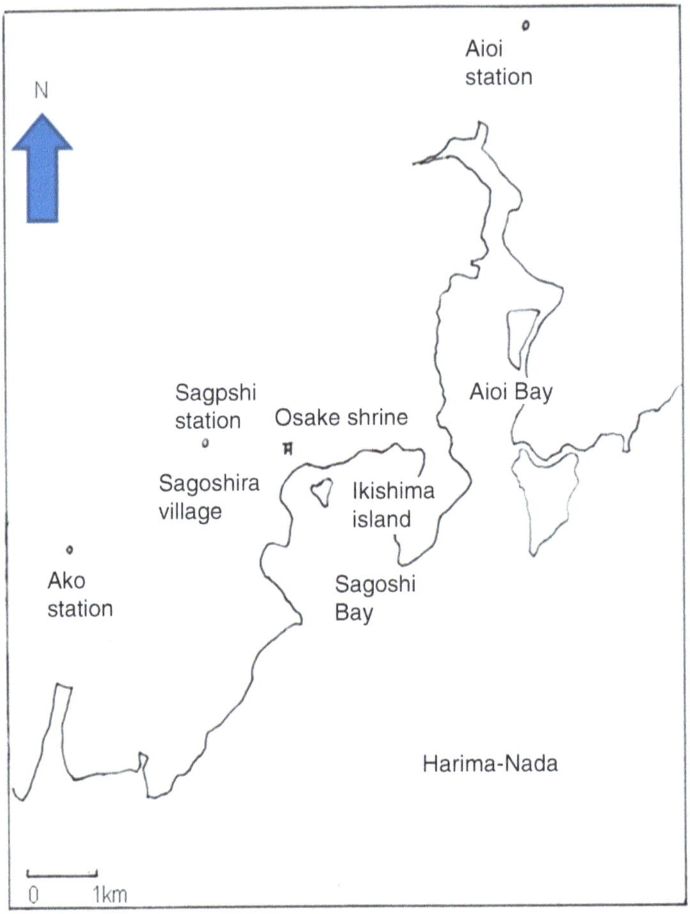

Fig. 3.40 Sagoshi Bay and Ikishima

Similarly, we have Chinjyu-no-Umi (God's Sea) where the local people have not touched the biota; for example, the coastal sea around Ikishima Island at Sagoshi Bay in the western part of Harima-Nada, the Seto Inland Sea. The history of Chinjyu-no-Umi at Sagoshi Bay is introduced in this section.

3.10.1 Ikishima Island

Sagoshi Bay is a small bay located in the northwestern part of Harima-Nada, Seto Inland Sea, and Ikishima Island exists at the head of Sagoshi Bay. The whole area of Ikishima Island belongs to the Osake Shrine, which is located at Sagoshi-ura village at the front of Ikishima Island (Fig. 3.40).

Fig. 3.41 Tomb of Mr. Kawakatsu Hata on Ikishima Island

The surrounding length (circumference) of Ikishima Island is 1.63 km, and the trees in the island are preserved in a pristine state because landing of people on the Island is prohibited. Most of the trees on the island are pasania, which were designated as a National Treasure in 1924.

The Osake Shrine worships God Osake (Mr. Kawakatsu Hata), God Amaterasu, and God Kasuga. The year that this shrine was built is not clear, but the shrine was made very famous in the 1180s by the book "History of Shrines in Harima."

The legend says that Mr. Kawakatsu Hata, who was the brain of Shotoku-Taishi in the Asuka era (592–710), was attacked by Mr. Iruka Soga after the death of Shotoku-Taishi in the early 600s and that he escaped from the capital of Yamato by a ship which was headed to the west in the Seto Inland Sea. His ship was wrecked off Sagoshi Bay but he arrived safely at Ikishima Island. He lived on this island and worked for the development of Sagoshi-ura village. Natural disasters continuously happened after his death in 647. Therefore, the local people built the tomb of Mr. Kawakatsu Hata in Ikishima Island (Fig. 3.41) and the shrine in Sagoshi-ura village to prevent the occurrence of natural disasters, and people have been prohibited from entering Ikishima Island (Fig. 3.42).

Another legend says that some staff of Ako castle cut the trees in Ikishima Island for building of the castle in the Edo era (1603–1868) and their master, Mr. Asano-Takumino-Kami, was killed. Therefore the local people do not want to enter into Ikishima even now so as to avoid bad events.

Fig. 3.42 Ikishima viewed from Osake Shrine

3.10.2 Shipping at Sagoshi

Shipping by the Sagoshi-ura people was very popular, including Kitamae-Bune, until the middle of the Meiji era (1868–1912). Traders went out from Sagoshi port with salt, sold their cargo, and bought other goods in the western part of the Seto Inland Sea and the western part of Japan Sea, then came back to Sagoshi port with valuable stone. They donated the stone to the Osake Shrine, which is the guarding god of their sailing. The steps to the main building of Osake Shrine are made from seamless limestone blocks that are 6 m long (Fig. 3.43).

More than 40 pieces of art were donated to the Osaka Shrine, including a valuable ship drawing painted 280 years ago. We can understand that this shrine has been respected by the sailors. The autumn festival of this shrine called the "Ship-Festival at Sagoshi," which was begun in the early Edo era and has been taken care of by each area in Sagoshi-ura village in turn, is conducted on the second Sunday in October every year. It is one of the three largest ship festivals in the Seto Inland Sea (others are Tenjin in Osaka and Miyajima in Hiroshima) and was designated as a National Treasure in 1992 (Fig. 3.44).

The population of Sagoshi-ura village decreased after the middle Meiji era as a result of the decline of the shipping industry, and at present there are only about 500 houses in Sagoshi-ura village, although in the past there were more than 1,000. The main occupation of the people in Sagoshi-ura village is as workers in the small-scale factories nearby.

3.10.3 Fisheries in Sagoshi Bay

The fishermen in Sagoshi-ura, whose houses number only about 30 now, began oyster culture in the southern half of Sagoshi Bay 30 years ago. They think the surface water and groundwater, with rich nutrients from Ikishima Island with its

Fig. 3.43 Stone steps in front of the main building of Osake Shrine. Each step is made of one stone

rich flora, may be the source of the rich oysters in Sagoshi Bay. In other words, the trees on Ikishima Island are a Fish Gathering Forest (Uo-Tsuki-Rin in Japanese).

Uo-Tsuki-Rin has been very famous in Japan but the detailed mechanism of gathering fish by Uo-Tsuki-Rin is not clarified yet. Some chemicals and nutrients emitted from the rich flora on Ikishima Island may be useful for the growth of phytoplankton, and this may result in the gathering of zooplankton, small fish, large fish, and the many oysters, but the detailed mechanism is a study theme for the future.

The oysters harvested in Sagoshi Bay are sold at the local shops and used in the cuisine at local restaurants.

The spiritual background of the people who preserve Chinjyu-no-Mori may be the fear of bad events happening by cutting trees in Chinjyu-no-Mori. Therefore, we need some spirit of fear to conserve Chinjyu-no-Umi.

Fig. 3.44 Pictures and the ship for the festival in Osake Shrine

3.11 Seaweed Culture and Preservation of the Coral Reef: Onna Fishermen Union in Okinawa Prefecture

Onna town is located on the west coast in the northern part of Okinawa Island, Japan, with a population of 10,000 and a long coastline, 46 km in length (Fig. 3.45). The Onna Fishermen Union has 100 full members and 190 submembers (in March 2010), and the average age of members is as young as 49 years old.

The Onna Fishermen Union conducts not only fisheries resource management, such as the control of harvesting fish and bivalves, but also conservation of the marine environment, such as prevention of inflow of red soil to the coastal sea, catching devil-starfish (Onihitode in Japanese), which predate on the coral reefs, and conservation of the coral reefs (Yanaka 2009).

Such activities of the Onna Fishermen Union are introduced in this section.

3.11.1 Seaweed Culture

The main product of Onna Fishermen Union is 600 tons per year of cultured Mozuku (a kind of seaweed), which accounts for about 60% of the total production in Japan. They culture two kinds of Mozuku, that is, Okinawa-Mozuku (Hon-Mozuku) and

Fig. 3.45 Onna town in Okinawa Prefecture

Mozuku (Ito-Mozuku), and the production of Okinawa-Mozuku is more than 90% of the total production in Japan.

Onna Fishermen Union succeeded in Hon-Mozuku culture in 1977 for the first time in Okinawa Prefecture. They have tried the culture of Ito-Mozuku, which is more difficult to culture than Hon-Mozuku, since 1985.

However, ruined Mozuku was sent back to the Union in 1991 because the Union sent wet Mozuku (not well dried) to the consumer. After that event, the Union told its member to conduct stringent inspection of products, and the members claimed, in 1992, that (1) the Union must trust the members, (2) the members must trust the Union and try to sell the products by themselves, and (3) the Union may abandon the products if they cannot sell them.

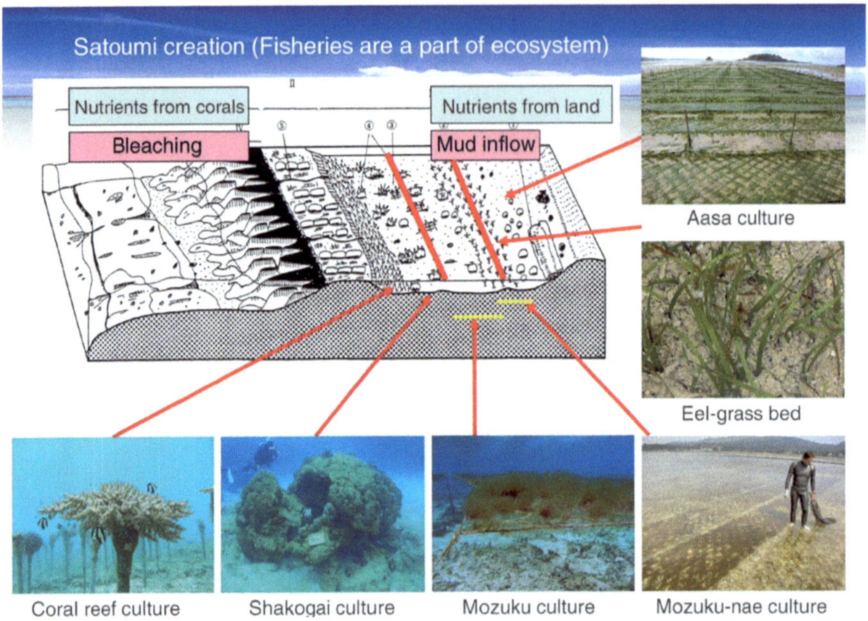

Fig. 3.46 Zoning of fishing ground by Onna Fishermen Union (Onna Fishermen Union 2008)

The management of Onna Fishermen Union is based on the product sections (e.g., Mozuku Section, Fish Section, Aasa Section), and the members have the responsibility of good management. The Mozuku made by Onna Fishermen Union has had the brand name "Made in Chura-Umi (beautiful sea)" since 2000. All the members feel that Mozuku are cultured well when the coral reefs are in good condition and that the Mozuku culture results in increasing the biodiversity of the coral reefs. Therefore, they have carried out conservation activity for the coral reefs such as coral culture and removing devil-starfish (Onihitode).

3.11.2 Zoning of Fishing Ground

The tidal flats, of 120 ha, called "Yakata-Katabaru," are used for "Aasa Culture (September–March)," "Mozuku-Nae Culture (December–March)," "Mozuku Culture (January–May)," and "Shako-Gai Culture (throughout the year)" from the coast, and its offshore area is occupied by coral reefs (Fig. 3.46).

The nets to which the seeds of Mozuku are attached in the water tank are spread in the Mozuku-Nae culture area in December.

Fig. 3.47 Mozuku-Nae (seaweed) culture

Aasa culture is conducted on nets, and Mozuku-Nae culture is conducted in a pond whose wall is made of stones (Figs. 3.47 and 3.48). This method is used because the nutrient concentrations in the shallow area are higher than offshore as a result of the nutrient supply from the land through surface water and groundwater, and the Mozuku-Nae needs many nutrients for its growth.

The biodiversity in Mozuku-Nae pond, which has water at all times, is higher than outside the pond, where there is no water at low water and no eelgrass bed, because a eelgrass bed has formed in the pond and small fish gather to the eelgrass bed (Fig. 3.49).

Mozuku grown at Mozuku-Nae Culture area in 1 month are moved to a place of Mozuku culture, and they are spread one-net by one-net, grown over the next 2 months, and harvested in April or May.

The committee for the prevention of inflow of red soil to the coastal sea is established in Onna town by the officers of the town, the fishermen union, workers for construction on land, and local residents, and it operates as the center of matters related to red soil inflow problems. Moreover, a Marine Protected Area is designated in Yakatakatabaru for the conservation of fish resources there.

Another threat to the fishing ground is the devil-starfish (Oni-Hitode), whose population explosions occurred in 1969, 1971, 1984, and 1986, and it destroyed the coral reefs. The reason of the population explosion of Oni-Hitode has not been

Fig. 3.48 Young Mozuku (seaweed) bed for culture

Fig. 3.49 Eelgrass bed formed in the cultured young seaweed bed

clarified. Local fishermen understand from their experiences that a catching rate of Oni-Hitode less than 45% is useless for coral reef preservation and that a catching rate of more than 60% of small-sized Oni-Hitode (less than 20 cm long) before brooding eggs is necessary to preserve the coral reef.

3.11.3 Collaboration for Conservation of Coral Reef

The Onna Fishermen Union acquired the fishing right of coral reef culture in 1998, established their coral reef culture section in 1999, and began the project of expanding the coral reef area after 2003. The coral reef culture section has carried out some related enterprises such as touring culturing coral reef juveniles, tour of planting coral reef, and tour of watching coral reef under cooperation with hotels, diving shops, and tour conductors for the promotion of this project.

The spirit of this project is written as "This project is done under the cooperation of all members of Onna Fishermen Union, the individual must do his best to cover his own responsibility and promote the coproduction with related other sectors" in the booklet of the Onnna Fishermen Union, "4th Innovation of Fisheries in Onna Area" (Onna Fishermen Union 2008).

The Onna Fishermen Union began cooperation with the Seikyo (COOP) and consumers such as (1) coral reef culture and expansion of coral reef area, (2) management of planted coral reef, (3) necessary projects such as information exchange, and (4) promotion of renewable fisheries products and direct sale of manufactured fishery products since 2008.

In Onna Fishermen Union, "coral transport" means collecting the pieces of natural corals and planting them at other places but "coral planting" means planting of the cultured corals.

The Mozuku sold by Seikyo includes a 1 yen donation per four pieces, and the potato ball made by Onna Fishermen Union sold in Okinawa includes a 2 yen donation per pack (six pieces) for coral reef conservation.

Such wide cooperation supports the coral reef conservation activity by Onna Fishermen Union.

3.12 Conservation of Fishery Resources in the Nearshore Region: Shiriya Fishermen Union in Aomori Prefecture

Shiriya Fishermen Union has conducted activities for the conservation of fish resources in its nearshore region (Fig. 3.50) and achieved valuable results.

Some activities of the Shiriya Fishermen Union are introduced in this section.

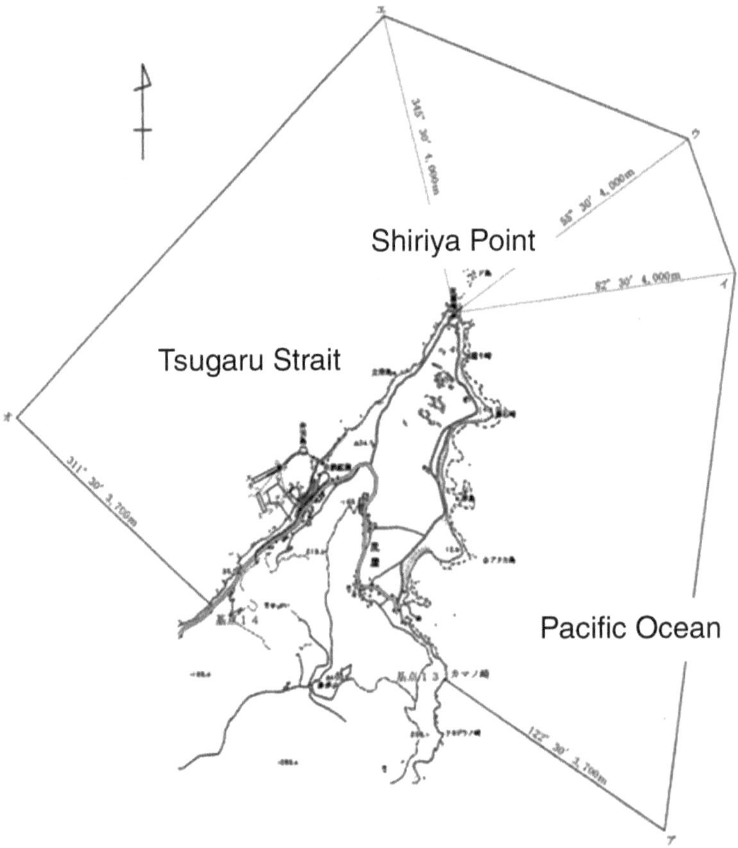

Fig. 3.50 Shiriya Point

3.12.1 Hamashiriya Relics

Higashidori Education Committee found an ancient abalone deposit place (shell midden) near Shiriya Village in 1995 (Fig. 3.51). At the same time, they found a kitchen from 600 years ago at the same place. Therefore, they thought that the abalone was cooked here 600 years ago. The Chinese pottery found there suggests that the trade between Japan and China was active at that time.

Such matters suggest that the people in Shiriya Village have carried out economic activities not only domestically but also internationally by using the coastal fishery resources such as abalone from 600 years ago.

Shiriya Point lighthouse was built in 1875, strong horses called Kandachime (strong for cold climate in Japanese) were released near Shiriya Point, and the area has become a sightseeing place (Fig. 3.52).

Fig. 3.51 Hamashiriya relics (now buried)

Fig. 3.52 Lighthouse at Shiriya Point and Kandachi horses

Shiriya village in Higashidori Town is a small village with 50 houses; most people in the village are fishermen, with 75 members in the Shiriya Fishermen Union with an average age of 50 years as of April 2010.

3.12.2 Fisheries of Shiriya Fishermen Union

The main fishing activities of Shiriya Fishermen Union are squid and octopus fishing offshore, seine net (throughout the year), tangle (kelp) gathering at the coast (April to October), Hunori (a kind of seaweed) gathering (April to June), sea urchins (April to July), and abalone (June to November) nearshore. Therefore, the nearshore fishing activities and the conservation of fish resources in the nearshore area are very important for the Shiriya Fishermen Union.

The fisheries production of Shiriya Fishermen Union has been stable at 0.6–0.8 billion yen per year, and seaweed, sea algae, and bivalves account for more than 50% of the production. The sale of the fisheries harvest is conducted by a professional merchant in Mutsu City near Higashi-domari town, and the members of Fishermen Union must pay 5% of their income to the Union.

3.12.3 Conservation Activities of the Marine Forest

The flora around Shiriya Point deteriorated in the Meiji era, the change to desert proceeded, and the coastal sea resources were greatly damaged by sand deposition from the air. Then the people in Shriya Village began planting the black pine tree in 1911 and restored the fish-gathering forest in 1940, 40 years after the beginning (Fig. 3.53a, b).

The tangle (kelp) forest in Shiriya coastal sea disappeared because of volcanic ash from Komagatake in 1929. The fishermen in Shiriya Village began to reproduce the Konbu forest in 1963, developed the tangle forest reproduction method using rope lines (Fig. 3.54) after trial and error, and succeeded in the reproduction of 300 ha of tangle (kelp) forest in 1982. After that time, the direct harvest of tangle from the marine forest was prohibited, and only picking up the drifted tangle along the coast is permitted (Fig. 3.55). As a result, the harvest of tangle has been steady at about 200 tons (0.2 billion yen) per year (Fig. 3.56).

Shiriya Fishermen Union established the Shiriya Fisheries Research Group in 1963, continued coastal sea monitoring (water temperature, transparency, number and weight of tangle, individual numbers, length, diameter, and weight of sea urchin and abalone, and so on) in June every year. Monitoring is carried out at 70 points (every 5 points along 14 lines; Fig. 3.57). The staff of Aomori Prefecture Fisheries Experimental Station cooperates with the members of 20 fishermen, and a total of seven fishing boats join this monitoring. Shiriya Fishermen Union conducts tangle harvest forecasting using the results of this monitoring; that is, they say that the tangle harvest will increase in years of low water temperature in June.

The Fisheries Research Group has carried out the transferring of sea urchins to reduce the grazing pressure of sea urchins on the tangle forest since 2003 (Fig. 3.58).

Fig. 3.53 (**a**) Rehabilitated forest for fish near Shiriya Point (photograph kindly provided by Mr. M. Inui). (**b**) Pine trees in the forest for the benefit of fish

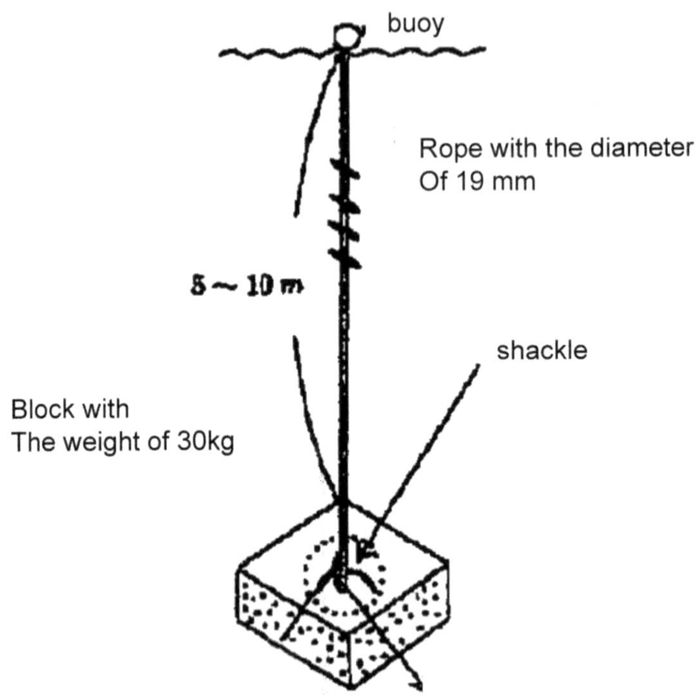

Fig. 3.54 Device with rope for attaching tangle (kelp sporelings)

Such activities of conservation of fisheries resources in the nearshore region of Shiriya Fishermen Union have continued for 100 years. The main reason of such continued activities for a long time are considered to be (1) the existence of a local philosophy called Sanyo (which is explained in the next section), (2) the hegemony of young fishermen (16–42 years old) who belong to the Sanyo association, (3) the fisheries production depending on the narrow sea algae beds along only 6 km of the long coastline, and (4) the sense of commons of village people (37 houses) regarding coastal resources.

3.12.4 Sanyo Association

The Sanyo Association was established as a local organization in 1911, and its fundamental spirit is (1) the power remaining from summer exists in winter, the power remaining from daytime exists at night, and the power from a fine day exists in rainy

Fig. 3.55 Women of Shiriya Fishermen Union collecting drifted tangle

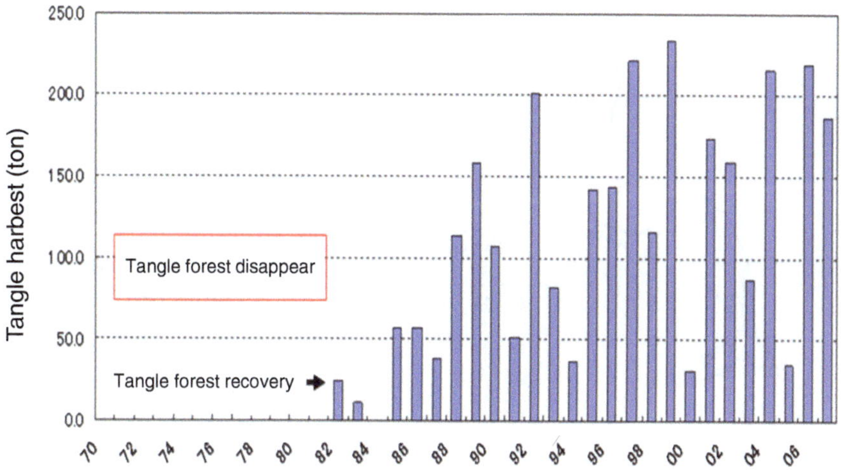

Fig. 3.56 Year-to-year variation in tangle harvest of Shiriya Fishermen Union (kindly provided by Mr. M. Inui)

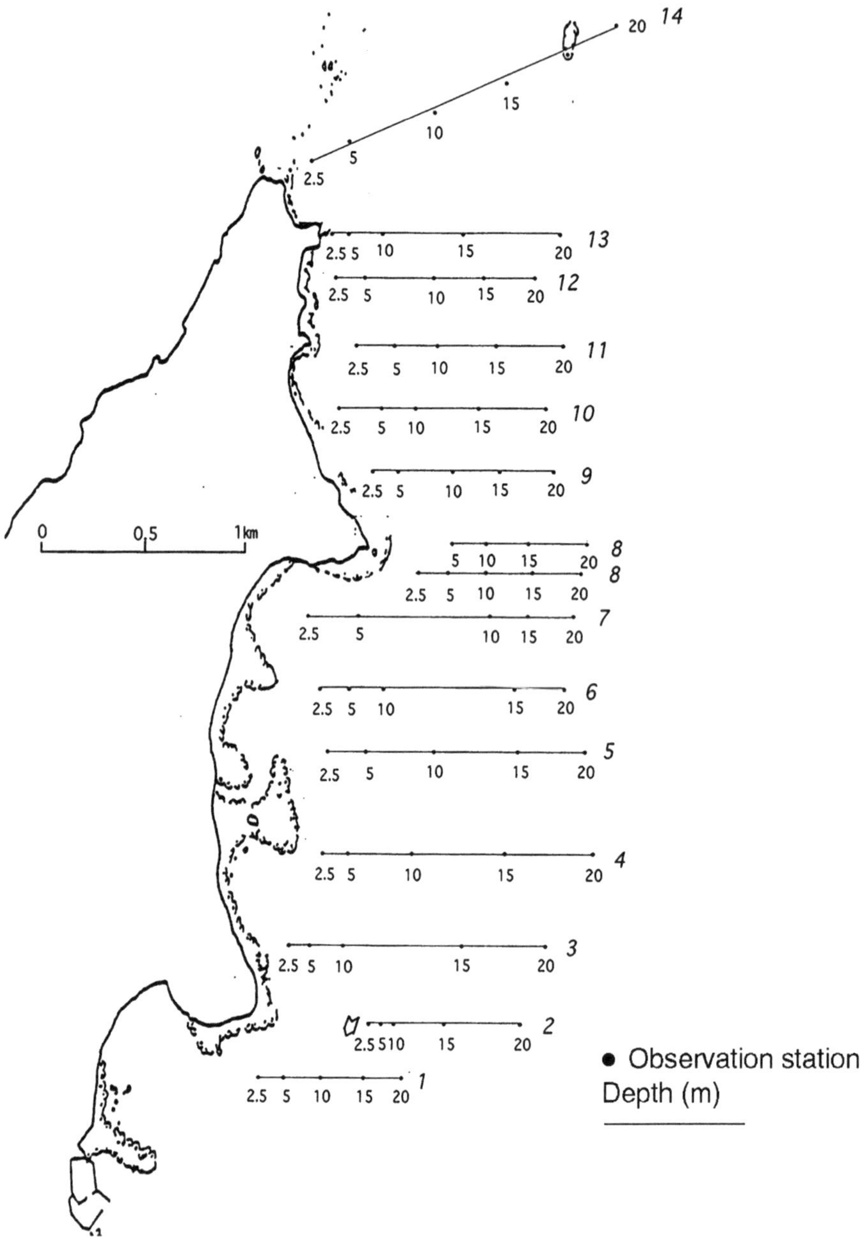

Fig. 3.57 Monitoring stations by Shiriya Fisheries Research Association

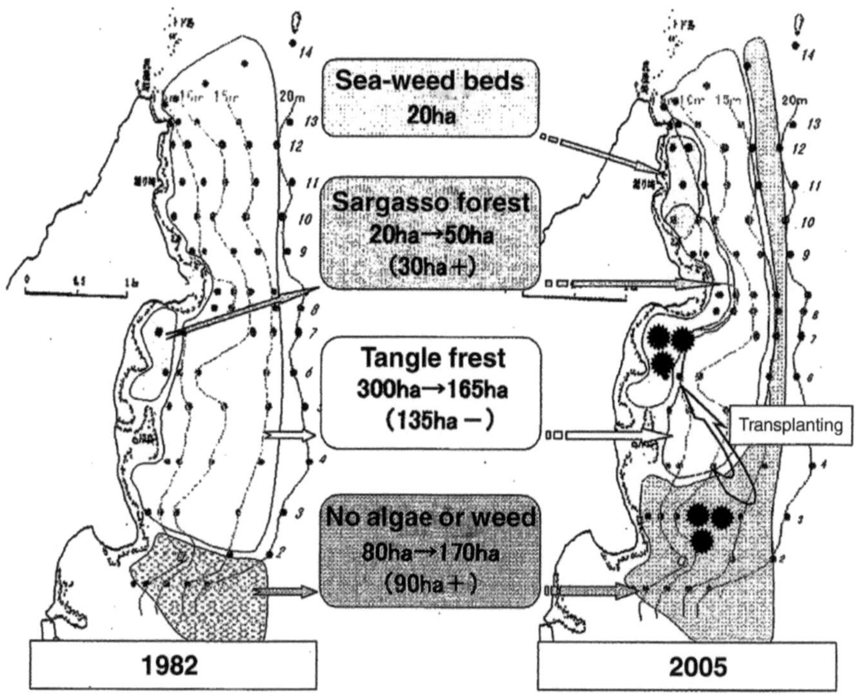

Fig. 3.58 Moving sea urchins to prevent the disappearance of the marine forest (kindly provided by Mr. M. Inui)

days, and (2) therefore young men must study in winter, on rainy days, and at night by using such remaining power. This thought was originally developed by the late local leader Mr. Musashi Hayasi. The members of this association must be young men whose age over 16 and under 42 in Shiriya Village; at present there are 22 members (in 2010), and the objectives of the association are (1) development of local industries, (2) exchange of information, (3) social education, and (4) improvement of daily behavior.

Moreover, the philosophy of pristine communism, that is, "get the products by cooperation and distribute them to everyone in the same weight," has been transmitted from olden times to the present time in Shiriya Village.

Chapter 4
Overseas Dissemination

4.1 8th EMECS

The 8th EMECS (Environmental Management of Enclosed Coastal Seas) was carried out at the International Large Light Hotel in Shanghai, China, from 26 to 30 October 2008 with 450 participants from 36 countries.

On 26 October (Monday), 11 topics were presented in the preliminary session after the opening ceremony, which were mainly on the EBM (Ecosystem-Based Management) in the Changjiang Estuary, Delaware Bay, Liverpool Bay, and so on. It was a surprising change because the main monitoring items were water quality such as COD (chemical oxygen demand), TN (total nitrogen), TP (total phosphorus), DO (dissolved oxygen), and so on at the time of the 7th EMECS in 2006 at Caen, France. It was reported that the hypoxic areas decreased in the northwestern part of the Black Sea and that the swimming crab catch increased in Delaware Bay as a result of ECM.

4.1.1 Sessions

On 27 October (Tuesday), the sessions of "Environmental change in the catchment area and estuary due to the global warming," "Policies on environmental and social risks related to integrated environmental management," "Geological change and disaster assessment in mega-delta," and "Environmental education for the youth" were held. It was interesting that intermittent outflow from the dam resulted in the increase of biodiversity in the estuary.

On 28 October (Wednesday), the sessions of "Water quality in the large river and its relation to the land-use change in the catchment area," "Total load control in the coastal seas in the world," "Cooperation between the coastal sea sciences and the coastal zone management," and "Satoumi" were held.

T. Yanagi, *Japanese Commons in the Coastal Seas: How the Satoumi Concept Harmonizes Human Activity in Coastal Seas with High Productivity and Diversity*, DOI 10.1007/978-4-431-54100-4_4, © Springer Japan 2013

4.1.2 *Satoumi Section*

In the "Satoumi" session, examples of the relationship between coastal zone management and the Satoumi concept on a worldwide scale, such as "Satoumi concept is useful for the compromise among the different stakeholders related to the oyster in Chesapeake Bay, USA," "EBM (Ecosystem-Based Management) in EU (European Union) concept is similar to Satoumi concept," "The role of scientists in the total load control of TN (total nitrogen) and TP (total phosphorus) in Masan Bay, Korea was great," "IMTA (Integrated Multi-Trophic Aquaculture), where fish, sea-weed and sea-cucumber are cultured in the same area and this is zero-emission culture, is very successful in China," "Waste water in the shrimp pond is cleaned by the oyster cultured in the new waterway built between the shrimp pond and the mangrove area and the sustainable shrimp and oyster culture method is established," "Sasi is respected in the local village but sasi is destroyed by the people invaded from the other village in Indonesia," and "ICZM (Integrated Coastal Zone Management) by citizens, scientists, and local government results in the recruitment of short-necked clam in tidal flats in the western Japan," were introduced.

In the general discussion, the participants agreed that "Not only the scientific knowledge but also the information on the local history, culture and religion and the marketing of fishery products are necessary to sustain the coastal fishery resources and to disseminate satoumi concept in the world." The details of this session are referred to in the report from the International EMECS center (Fig. 4.1; International EMECS Center 2009).

On 29 October (Thursday), at the last plenary session, where the results of sessions were reported, the participants adopted the Shanghai Declaration of "The world suffers from the big economic crisis now but we have to challenge the successful integrated coastal management in the coastal sea and catchment area for the conservation of coastal resources based on the new concept of Satoumi because the coastal resources will give the economic benefit to human beings."

4.2 3rd EAS

The 3rd EAS (East Asia Seas) by PEMSEA (Partnerships in Environmental Management for the Seas of East Asia) and the Philippines Government was held from 23 to 26 November 2009 at the Philippines International Convention Center with 1,500 participants from 20 countries.

This conference was the third one, following the first one at Putrajaya, Malaysia in 2003 and the second one at Hainan Island, China in 2006; 34 sessions related to ICM (Integrated Coastal Zone Management) were held during 4 days of the conference, and the "Satoumi Session" was held on 24 November.

Fig. 4.1 Report of Satoumi section at Shanghai

4.2.1 Opening Ceremony

The keynote lecture by Mr. Kei-ichi Katayama (Kaiyo Construction Corp., Japan) in the opening ceremony on 23 November (simultaneous translation to English was conducted on the screen) was very impressive because he claimed that "I have

Fig. 4.2 Report of Satoumi session at Manila

developed the artificial reef made by bivalve's shell called 'Shell-nurse' in these 20 years based on the long experience of fishermen, that is, much fish gather to oyster mound. I employ free fishermen in the season without fishing activity for the construction of artificial reef of abandoned shell. The biodiversity and fish resources are increased by the injection of such artificial reef in the Japanese coastal seas. This is a kind of sustainable coastal fisheries."

4.2.2 Satoumi Session

About 70 people from ten countries attended the "Satoumi Session" (Chair, Dr. O. Matsuda; Vice Chairs, Dr. T. Yanagi and Dr. A. Maccdnard), which was supported by PEMSEA and the International EMECS (Environmental Management of Enclosed Coastal Seas) Center.

In the morning session, the definition of Satoumi was introduced, the difference between Western countries and Eastern countries on the relationship between humans and nature was discussed, and some examples of Satoumi creation in Japan such as Nanao Bay, Yamaguchi Bay, Ago Bay, Tokyo Bay, and Harima-nada, Seto Inland Sea were introduced.

In the afternoon session, "Recovery movement from the Indian Ocean big tsunami in Andaman Sea, Thailand," "Coral reef recovery movement with the citizen's cooperation in Thailand," "Conservation activity of fish resources based on the traditional religion in Bali Island, Indonesia," "Mangrove conservation activity by citizen in Vietnam," "Evaluation of the effect of artificial reef in Malaysia," "Conservation activity of tidal flats in Korea," "Fault and its reason of ICM in the Philippines," "Trial of changing the abandoned shrimp pond to mangrove forest in the Philippines," and "Conservation of fish resources based on the experience of fishermen" were presented.

In the panel discussion conducted at the end, the participants agreed that "It is possible to increase fish resources by construction of artificial habitats and the sustainable fishery is also possible by the suitable management of fish harvest. However, the construction of artificial habitats in the coastal sea is only possible under the suitable water quality which mainly depends on the water quality in rivers. Therefore the integrated management from mountain, land, river to coastal sea is very important to keep the good water quality in rivers, and the human production activity at each place, that is, forestry, agriculture and fisheries must co-exist with the nature there. For the success of satoumi creation, the successful interdisciplinary study among natural, social and cultural sciences is indispensable."

The details of this session are referred to in the report from the International EMECS Center (Fig. 4.2; International EMECS Center 2010).

Chapter 5
Conclusion

The theoretical consideration of the creation of Satoumi has been nearly finished, as discussed in Chap. 2. Some trials of creating Satoumi have already begun at some places in Japan, as introduced in Chap. 3. The remaining problem is the relationship between the fishing village and the city.

The main players in Satoumi are the fishermen who carry out production activities in the coastal sea, but the number of fishermen in Japan is less than 0.2% of the whole population, and some fishing villages are now disappearing.

We have to consider the sustainability of fishing villages by classifying Satoumi into three categories as follows:

1. Satoumi as fishing village: most of the village people are fishermen who receive money from fishing.
2. Satoumi near the city: the fishing village is near the city, so that city people can come there and return home within one day; the number of fishermen is less than half the population, and other people receive income by supporting the recreational activities of visiting city people.
3. Satoumi far from the city: the fishing village is far from a city; the number of fishermen is few, and other people receive income by supporting eco-tourism, excursions, and student tours from the city.

For point 1, it is possible to maintain such fishing villages as introduced in Chap. 3 under the appropriate guidance of a strong leader. Of course, many different ideas are necessary for a sustainable fishing village.

For point 2, some trials have been undertaken, as introduced in Sects. 3.6 and 3.7; the development of such a type of fishing village is not easy, and we have to endeavor to promote the co-production of fishing village and city.

As for item 3, some trials have been undertaken, such as at Miyajima in Hiroshima Prefecture and Ie-shima in Okinawa Prefecture, but the symbiosis of tourism and local industry is not so simple, as discussed in Sect. 3.9. More exchange between the fishing village and city under the carrying capacity of tourism is necessary.

T. Yanagi, *Japanese Commons in the Coastal Seas: How the Satoumi Concept Harmonizes Human Activity in Coastal Seas with High Productivity and Diversity*, DOI 10.1007/978-4-431-54100-4_5,

I am continuing the study of Satoumi, promoting the mutual exchange of fishermen and city people and clarifying desirable relationships of fishermen and city dwellers for the creation of Satoumi.

I shall be greatly pleased if you will give me your comments on this book.

Mid-autumn in 2011: Tetsuo Yanagi

tyanagi@riam.kyushu-u.ac.jp

References

Ashikaga Y (2006) Coastal preservation project based on the viewpoint of citizen. River 10:46–51 (in Japanese)

Connell H (1978) Diversity in tropical rain forests and coral reefs. Science 199:1302–1310

Fukuoka Prefecture (2008) Trend of fisheries in Fukuoka Prefecture, White paper of fisheries in 2007 (in Japanese)

Hagiwara T (2008) Actions for marine forest deterioration in Shonan district of Shizuoka Prefecture: references of symposium on symbiosis of human and nature. Japan Fisher Assoc 13–21 (in Japanese)

International EMECS Center (2009) International workshop on "Sato-Umi" at Shanghai Report

International EMECS Center (2010) International workshop on "Sato-Umi" at Manila Report

Itoh Y, Nakano K, Fujisawa S (2008) Artificial reef and small animals around it: fish resources increasing effect of artificial reef (1). Fisher Eng 45-2:101–110 (in Japanese)

Japan Fisheries Agency (2006) White paper of fisheries in 2005 (in Japanese)

Kakuma S (2009a) Multifunction of MPA in coral reef. In: Yamao M, Shima H (eds) Multifunction of fishing village and fisheries in Japan. Hokuto Press, Japan, pp 89–110 (in Japanese)

Kakuma S (2009b) Ecosystem conservation and fish resources management—example at coral reef of Yaeyama, Okinawa. Study Regional Fisher 49-3:67–89 (in Japanese)

Kamimura M (2007) Reproduction of rock weir at Shiraho, Ishigaki—coral reef conservation by local people. Regional Study 3:175–88 (in Japanese)

Kamimura M (2011) Possibility of biodiversity conservation using "Satoumi" as a key word—report of Shiraho's international stone tidal weir summit. Regional Study 8:17–28 (in Japanese)

Kokubu H, Takayama Y (2008) Characteristics of tidal flats and its reproduction. Ocean Biota 30-3:303–314 (in Japanese)

Kumamoto K (2000) What is wrong in public construction? Mana Press, 269pp (in Japanese)

Kuroshio Feeling Center (2006) Report of ecotourism of Kashiwajima in 2005, 126pp (in Japanese)

Larkin PA (1996) Concepts and issues in marine ecosystem management. Rev Fish Biol Fisher 6:139–164

Matsuda O (2007) Environmental restoration using multifunction of fisheries. In: Research Institute for Seto Inland Sea (ed) Satoumi creation in the Seto Inland Sea. Koseisha-Koseikaku, Tokyo, pp 29–38 (in Japanese)

Morimoto Y (2008) Conservation of biodiversity near our life. Book on green. Special issue of management of environment and resources, pp 2–10 (in Japanese)

Mosse JW (2008) Sasi Laut: History and its role of marine coastal resource management in Maluku archipelago. International workshop on "Sato-Umi" Report. International EMECS Center, Japan, pp 68–76

T. Yanagi, *Japanese Commons in the Coastal Seas: How the Satoumi Concept Harmonizes Human Activity in Coastal Seas with High Productivity and Diversity*, DOI 10.1007/978-4-431-54100-4, © Springer Japan 2013

Murai Y (1998) Sasi, Asia and marine world—wisdom and system to conserve the environment. Commons, Tokyo, 220pp (in Japanese)
Murota T, Mimata M (2004) Co-used forest and commons. Japan Opinion Coop, 280pp (in Japanese)
Nagao T (2005) Can the right of sea grass harvest stop the public construction project? World 2005–6, No.740, pp 265–272 (in Japanese)
Nakagawa K (2008) Establishment of local brand. Seto Inland Sea 55:42–45 (in Japanese)
Nakayama M (2006) Right to co-use the environment—a kind of environmental right. Seibundo, Tokyo (in Japanese)
Naruse K (1990) Reconstruction of eel-grass beds. Abstracts of 33rd youth presentation festival from the fishing villages of Okayama Prefecture, pp 10–23 (in Japanese)
Onna Fishermen Union (2008) Beautiful sea, Part 3, 54pp (in Japanese)
Pikitch EK, Santora C, Babcock EA, Bakun A, Bonfit R, Conover DO, Dayton P, Doukakis P, Fluharty D, Heneman B, Houde ED, Link J, Livingston PA, Manget M, McAllister MK, Pope J, Sainsbury KJ (2004) Ecosystem-based fishery management. Science 305:346–347
Preservation Association of the Seto Inland Sea (2007) Environmental Data Report, 207pp
Seino S (2008) Formation plan of coastal area: evaluation and utilization of changing land. J Jpn Real Estate Soc 22–1:52–60 (in Japanese)
Sekine M, Ukita M (2008) Rehabilitation of short-necked cram in Yamaguchi Bay. J Jpn Benthos Soc 62:88–92 (in Japanese)
Shono (2007) Creation of Local Brand. Seto Inland Sea 51:12–14 (in Japanese)
Takasaki H, Takakuwa M (1976) Japanese and beach. Japan Broadcast Association Press, Tokyo (in Japanese)
Takatani K (2007) Beach corridor project in the Seto Inland Sea. Seto Inland Sea 49:47–51 (in Japanese)
Tanimoto (2010) Biodiversity in sea-glass bed and human interaction. Coastal Oceanography in Japan, 32:201–210 (in Japanese)
Tawa M (ed) (2006) Rock weir: oldest fishing trap. Hosei University Press, Tokyo (in Japanese)
Tsunematsu A, Kato M (2008) Environmental change in artificial tidal flat. Abstracts of civil engineering reports in Central Area Branch, Japan, Sec.VII, 1–6 (in Japanese)
Ukita M (2007) Restoration of nature by connecting forest, river and sea—trial in Fushino River catchment. In: Research Institute for Seto Inland Sea (ed) Satoumi creation in the Seto Inland Sea. Koseisha Koseikaku, Tokyo, pp 51–66, in Japanese
Umino S (2008) Management for marine forest deterioration in Shonan Area in Shizuoka Prefecture. In: 7th technical reports on fishing ports and grounds, pp 31–36 (in Japanese)
Yanagi T (1998a) Civil engineering and environmental conservation. J Jpn Civil Eng Soc 83–11:32–33 (in Japanese)
Yanagi T (1998b) Satoumi creation in the coastal sea. J Jpn Water Environ Soc 21:703 (in Japanese)
Yanagi T (2004) Fisheries resources management at Himeshima. Engineering Sciences Reports, Kyushu University, 26, pp 215–217 (in Japanese)
Yanagi T (2005a) Seto Inland Sea—Introduction to Satoumi. Association of Seto Inland Sea Preservation, Kobe, 69pp (in Japanese)
Yanagi T (2005b) Fishery resources management of Misaki Fishermen Union. Engineering Sciences, Kyushu University, 26, pp 431–433 (in Japanese)
Yanagi T (2006) Satoumi. Koseisha-Koseikaku, Tokyo, 102pp (in Japanese)
Yanagi T (2007a) Sato-Umi: a new concept for coastal sea management. TERRAPUB, Tokyo
Yanagi T (2007b) Sale strategy for supporting Satoumi—examples of Misaki and Himeshima Fishermen Unions. Engineering Sciences, Kyushu University, 29, pp 315–317 (in Japanese)
Yanagi T (2008a) Traditional laws related to satoumi. Reports of Research Institute for Applied Mechanics, Kyushu University, 135, pp 105–108 (in Japanese)
Yanagi T (2008b) Style of satoumi study—interaction of science and society. Reports of Research Institute for Applied Mechanics, Kyushu University, 135, pp 109–1112 (in Japanese)

Yanagi T (2009a) Drifting gavage on the sea surface and depositing one in the Seto Inland Sea. Seto Inland Sea 59:4–7 (in Japanese)
Yanagi T (2009b) Biodiversity and human interaction. Jpn Oceanogr 18:393–398 (in Japanese)
Yanagi T, Ishii D (2008) Hypoxia in the Seto Inland Sea. In: Yanagi T (ed) Sea bed environment of the Seto Inland Sea. Koseisha Koseikaku, Tokyo, pp 77–88 (in Japanese)
Yanaka S (2009) Development of multifunction of fishing village: viewpoint from Satoumi creation. Study on Regional Fisheries 49:111–126 (in Japanese)
Yanagi (2010) Oceanography and Society. Oceanography in Japan, 19:243–248 (in Japanese)

MIX
Papier aus verantwortungsvollen Quellen
Paper from responsible sources
FSC® C105338

If you have any concerns about our products, you can contact us on
ProductSafety@springernature.com

In case Publisher is established outside the EU, the EU authorized representative is:
Springer Nature Customer Service Center GmbH
Europaplatz 3, 69115 Heidelberg, Germany

Printed by Libri Plureos GmbH
in Hamburg, Germany